HANDS-ON VHDL

AN INTRODUCTION USING REAL HARDWARE

HANDS-ON VHDL

AN INTRODUCTION USING REAL HARDWARE

Blaine C. Readler

Full Arc

HANDS-ON VHDL
AN INTRODUCTION USING REAL HARDWARE

Published by Full Arc Press

Visit us at: http://www.readler.com

E-mail: blaine@readler.com

ISBN: 978-0-9992296-9-9

Printed in the United States of America

First Edition: 2024

Training is everything. The peach was once a bitter almond;
cauliflower is nothing but cabbage with a college education.
— Mark Twain

Contents

Chapter 1

Preliminaries

A note before starting: to get even a minimal amount of value from this book, you will need:

o a computer running Windows 10 or later with at least one free USB port and some GBs of disk space;

o to purchase a development board (explained below) for between $150 and $165;

o to download free FPGA development software from one of the FPGA vendors.

Note that all source code files developed in the course of this book are available for download on my website:

http://www.readler.com – Engineering/Technical
References

As we well know, the world of electronics is divided into hardware and software. Hardware, we can touch – the wires and connectors and circuit boards mounted with IC components. Software is ethereal, pages of code that have been compiled into billions of bits of data instructions that the hardware executes one at a time. Software is a logical process: if this thing is greater than that and less than this other thing, then do something.

Hardware can perform logic operations as well. In fact, before there ever was software, hardware performed rudimentary logical operations with individual logic gates (ANDs, ORs, XORs, etc.) each built up from a handful of transistors. That said, it would have required an entire circuit board the size of a book to perform what a single line of code today can achieve.

Between the worlds of circuit board hardware and software code, bridging the dynamic adaptability of software and the blinding speed of hardware, lies a type of component that operates via billions of individual logic gates, all able to do their tiny piece of the job simultaneously, but whose interconnection – what comprises their final operation – is designed using specialized languages, code that compiles like software. These hybrid hardware/software IC components are called FPGAs (Field Programmable Gate Arrays), and the code languages are almost exclusively either VHDL or verilog.

By the way, a moderate sized FPGA now contains as many logic gates as a million of those early discrete component circuit boards.

In this book, we learn basic digital design using an FPGA as coded with the VHDL language. We develop a working knowledge of digital design techniques by implementing increasingly complex projects on either one of two introductory FPGA development boards sold by Terasic (Intel/Altera FPGA) or Digilent (AMD/Xilinx FPGA). These are the boards, one of which you will need to execute the projects of the book:

Terasic: D10-Lite board, Intel/Altera MAX10 FPGA, ~$140;
Digilent: Basys 3 board, AMD/Xilinx Artix-7 FPGA, ~$165.

Both offer significant discounts for students.

There already exists a multitude of project examples online for these and other FPGA development boards, but they, to the last one, assume that you are already proficient with VHDL and digital logic design. We, on the other hand, use a series of projects as a teaching tool, each project advancing your comprehension.

The following projects do assume, however, that you have a rudimentary understanding of logic gates, binary and hexadecimal numbers, and basic clocked register operation. If not, take a few hours to run through the free Youtube-based tutorial class in order to come up to speed:

U of Blaine presents DIGITAL DESIGN.

Simulation is an important tool when designing complex logic, and free limited-capability simulation software is provided free by the FPGA vendors. We do not cover simulation here, however, since the combination of fairly simple designs that compile in just a few minutes, and the ability to easily observe results (e.g. via the development board's LEDs and displays) eliminates to a large extent the need for simulation.

For each project, we review the operation with block and some timing diagrams before diving into the VHDL code. Each project allows some amount of "play," whereby you can experiment and explore detailed aspects of each subsequent newly presented material.

The first step is to go to either the AMD/Xilinx or Intel/Altera website (depending on which board you are purchasing) and download their free version of development software. Each vendor offers a purchased version of software for their medium-to-large FPGAs, but provides free versions for the smaller FPGAs:

Intel/Altera: Quartus II, Lite Edition;

AMD/Xilinx: Vivado, Web Version.

You can select the latest version, but probably any version within the last few years will work for these development boards.

The next step is to set up a folder area on your computer. There are many ways to structure your projects, and this is my standard method.

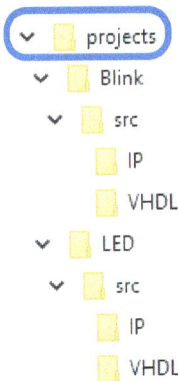

At the top, I create a comprehensive folder, here called "projects." Under this I place all the individual projects that we would be exploring. Here I'm showing the first two projects, "LED" and "Blink":

```
v  📁 projects
   v  📁 Blink
      v  📁 src
         📄 IP
         📄 VHDL
   v  📁 LED
      v  📁 src
         📄 IP
         📄 VHDL
```

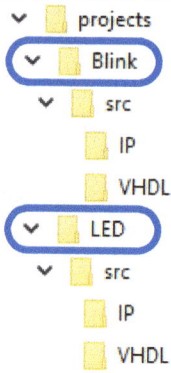

Within each project folder I create a sub-folder that I call "src". This is where the various source files for this project live:

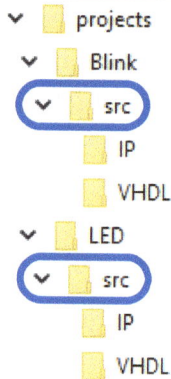

```
v  📁 projects
   v  📁 Blink
      v  📁 src
         📄 IP
         📄 VHDL
   v  📁 LED
      v  📁 src
         📄 IP
         📄 VHDL
```

The key source files are the design files that we create using the VHDL language, and I locate these in the "VHDL" folder. The important thing to remember here is that these VHDL design files are universal, and not tied to any one vendor or FPGA type:

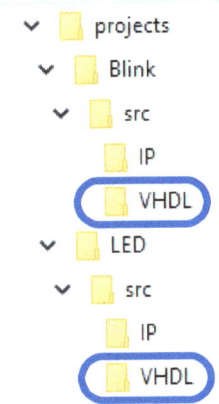

The parallel "IP" folder is where I keep those source files that are specific to one vendor, sometimes created within the vendor tool and configured specifically for this project's design application (e.g., the depth and width of a memory).

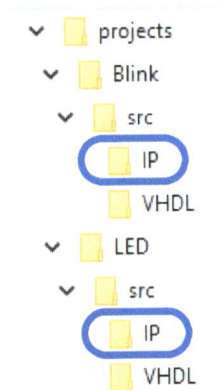

That described my standard folder layout, but in this book we are, of course, dealing with two vendors, and so the folder structure is modified a little to accommodate this. Normally, the vendor's project files – those that are dedicated to this particular vendor – reside at the top of each project (parallel with the "src" folder). However, since we have two vendors, I am creating a sub-folder for each vendor:

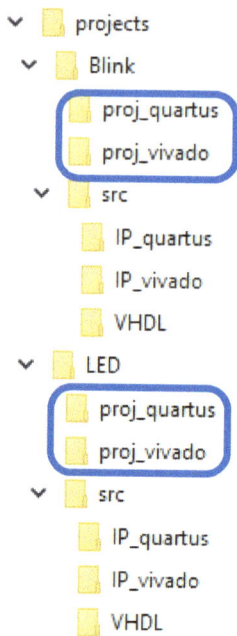

Additionally, I use a dedicated "IP" folder for each vendor.

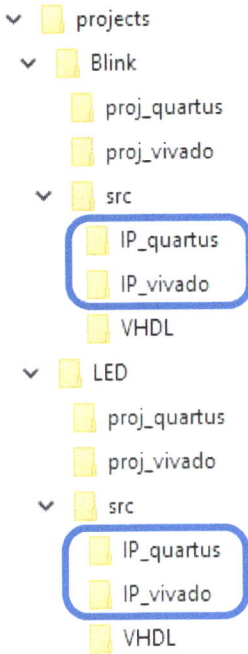

At this point you'll need to step aside and download and install whichever vendor development tool that you'll need for the board you've chosen: Intel/Altera's Quartus for the Terasic board, or AMD/Xilinx's Vivado for the Digilent board. This may require some patience, a fair amount of available computer memory, and a high bandwidth internet connection (gigabytes of download), particularly for Vivado. Remember that you're after their free version, which should not require a license. As their website methods change with time, any instruction here would soon be obsolete and potentially confusing.

A note about punctuation: commas and periods are generally placed before closing parenthesis. For example, the following words might describe my approach to writing this book: "fastidious," "thorough," and "clarity." However, I have taken the liberty to break this rule in order to avoid confusion about the exact spelling of signal names. So, for example, in this context I might write that "in_1", "out_1", and "enable_b" comprise all the signals of block "mux_2".

Good luck.

Blaine C. Readler

Chapter 2

LED

Our first project will serve to get us up and running with the rudiments of VHDL and the vendor software. In other words, don't be put off by the simplicity. Initially all we'll do is light an LED. However, getting that will mean that we've set the stage for all that is to come.

Here's our logic to light an LED.

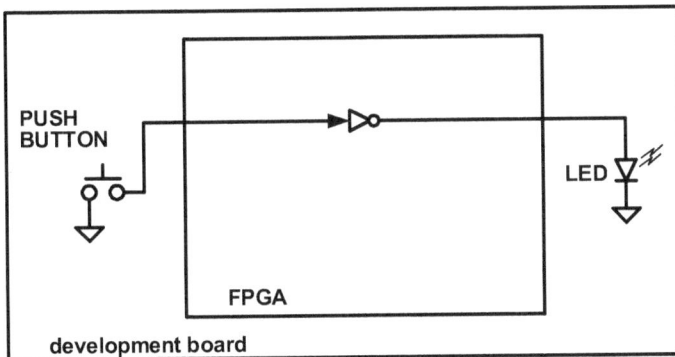

As simple as can be. While we push a button on the board, an LED either lights (for the Terasic board) or turns off (for Digilent the board). The push-button on the Terasic board connects the external signal to ground, and the FPGA must invert the input pin to light the external LED with a high level (and this turns off the LED on the Digilent board).

VHDL Source

The next step is to create our top-level VHDL file. These initial projects are so simple that we only need the top-level VHDL file – the entire design is included here.

The VHDL language, like any language, has specific requirements regarding its structure, key words, and syntax. We'll introduce each detail as needed as we implement the increasingly complex projects.

All VHDL files have the following parts.

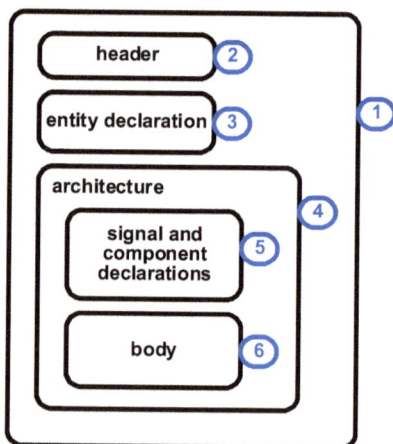

1) the file itself;

2) a header – notes, etc. that are not included in the compiling;

3) entity declaration – defines the IO (input/output) signals for this file;

4) architecture – the meat of the design. Complex projects can have multiple architectures, but we don't;

5) signal and component declarations – defines the signals and components (i.e., hierarchical sub-files) used in the body design;

6) the body, which contains all the VHDL that defines the design of this file.

The next figure shows the actual VHDL code for this project.

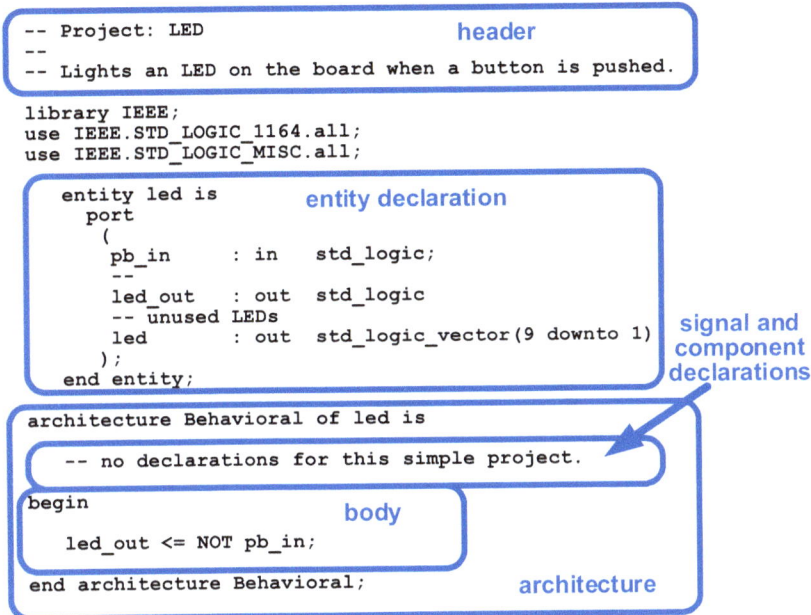

```
-- Project: LED                        header
--
-- Lights an LED on the board when a button is pushed.

library IEEE;
use IEEE.STD_LOGIC_1164.all;
use IEEE.STD_LOGIC_MISC.all;

  entity led is        entity declaration
    port
      (
        pb_in     : in    std_logic;
        --
        led_out   : out   std_logic
        -- unused LEDs
        led       : out   std_logic_vector(9 downto 1)    signal and
      );                                                   component
    end entity;                                            declarations

architecture Behavioral of led is

    -- no declarations for this simple project.

begin                    body

    led_out <= NOT pb_in;

end architecture Behavioral;                architecture
```

We'll look at the details of each section in turn, and we begin with the header. The double-dashes alert the compiler tool that what follows should be ignored, that this is merely comments, and not part of the actual VHDL code.

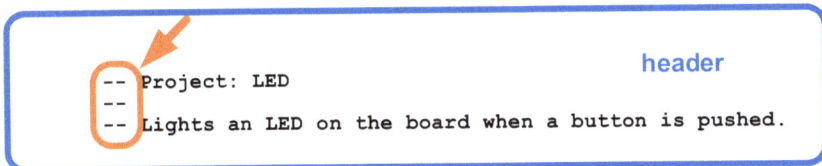

```
                                              header
    -- Project: LED
    --
    -- Lights an LED on the board when a button is pushed.
```

Next is the entity declaration. Everything shown in red is required, and will be found in every entity declaration, including the parenthesis, colons, and semi-colons.

Blaine C. Readler

```
entity led is                          entity declaration
  port
    (
    pb_in        : in    std_logic;
    --
    led_out      : out   std_logic
    -- unused LEDs
    led          : out   std_logic_vector(9 downto 1)
    );
end entity;
```

These are the parts that are specific to our design.

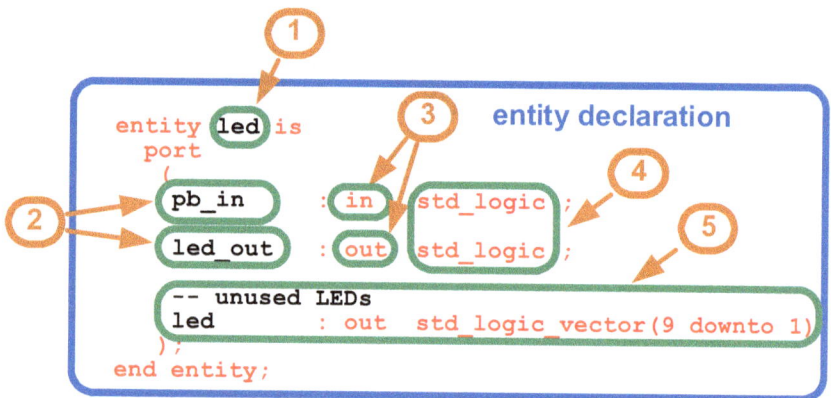

1) the entity name. This is the name that the compiler tool associates with the contents of this file, i.e., this part of our design (in this case, the only part). Although VHDL does not require that the entity name match the file name, I highly recommend that you keep them always the same;

2) the entity entry/exit signals. Since we have just one entity (this file) for this design, these are the names of the signals that enter and exit the FPGA on the circuit board, i.e., are the top-level signals;

3) port directions. The compiler needs to know if the signal is an input or output, using the keywords "in" and "out";

4) VHDL allows a variety of signal types, and we need to tell the compiler which types these port signals are. The great majority of signal types are "standard logic," and we indicate this with the "std_logic" keyword;

14

5) we'll ignore this line for now – it's necessary for the unused LEDs on the board, and we'll find it included in every project.

Take note of the following key parts, details that are often missed, and can create head-scratching when the compiler issues an enigmatic error, often not even telling you that something is missing.

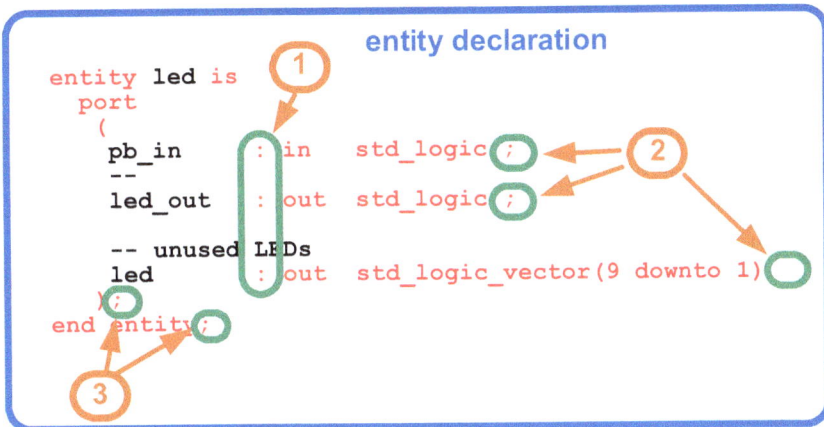

1) a colon separates every port signal from its direction indication;

2) a semi-colon marks the end of every port declaration except the last one;

3) semi-colons mark the end of the port list, and the entity declaration itself.

Finally, the architecture. Again, everything shown in red is required, and will be found in every entity architecture.

```
architecture Behavioral of led is

    -- no declarations for this simple project.

begin

    led_out <= NOT pb_in;

end architecture Behavioral;
                                      architecture
```

And, the parts specific to our design.

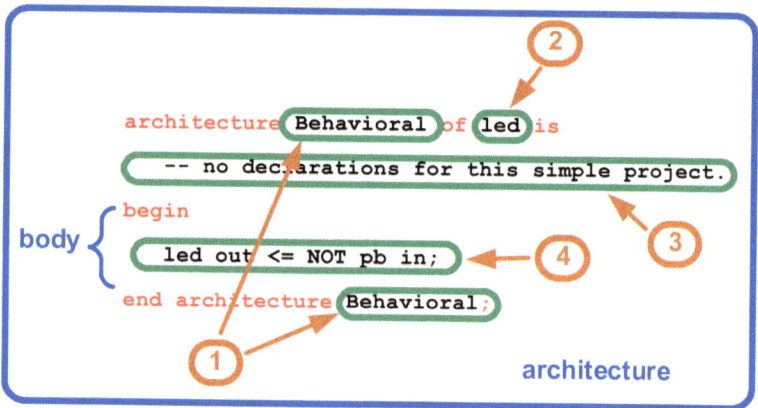

1) we can have different forms of an architecture, and the word "behavioral" describes the most common type. That said, in this case, "behavioral" is the name of our architecture, and could be anything we like;

2) "led," on the other hand, refers specifically to the entity to which this architecture belongs. In all our designs, we will have just one architecture, and it always belongs to the entity defined in the same file;

3) as we saw, this simple project has no signals to declare;

4) finally, the meat, which consists of one inverter;

Let's take a look, though, at that inverter, and it's connections to the input and output signals.

led_out <= NOT pb_in ;

1) the less-than/equal combination indicates an assignment – the result of logical operations on the right side is assigned to the signal on the left. So this line says, "the output, led_out, is the inversion of pb_in";

2) NOT is VHDL for an inversion;

3) assignment statements always end with a semi-colon.

A further note about that "NOT." I've made it upper case just to differentiate the VHDL operator from the signal names, however VHDL is "case insensitive," meaning it doesn't care whether the name, or even individual letters of the name, are capitalized or not.

However . . . although VHDL is also case-insensitive to signal names as well, I highly urge you to be consistent. Although VHDL doesn't care, the verilog language does, and someday down the road, should somebody (maybe you!) try to port the design to verilog, mixed cases could cause all sorts of havoc, possibly insanity.

And a word about file extensions, although some compile and simulation software may not care, the standard is to use ".vhd" (or ".vhdl").

Before we leave the VHDL source file, though, we need to address the pesky few lines regarding libraries.

```
-- Project: LED
--
-- Lights an LED on the board when a button is pushed.
library IEEE;
use IEEE.STD_LOGIC_1164.all;
use IEEE.STD_LOGIC_MISC.all;

    entity led is
      port
        (
        pb_in      : in    std_logic;
        --
        led_out    : out   std_logic
        -- unused LEDs
        led        : out   std_logic_vector(9 downto 1)
        );
    end entity;

architecture Behavioral of led is

    -- no declarations for this simple project.

begin

    led_out <= NOT pb_in;

end architecture Behavioral;
```

Like software programming languages, VHDL relies on pre-made libraries for support. For now, suffice it to say that these couple of libraries, which are included in all vendor software, comprise the minimal needed for very basic logic (like this first design). As we progress through more complex operations, we'll find that we will be adding additional libraries for support (all also included in the vendor software).

The Terasic/Quartus project

You have presumably installed Quartus, and are now ready to create and compile a project for our first simple design. There's one additional step before we do, however. We have to tell Quartus what FPGA pins we want to use for the external signals. We do this with what we call a "constraints" file. The first time you compile your project, Quartus creates a [].qsf file, where [] is the name of the Quartus project. To this, we add the pin assignments (and recompile), like so:

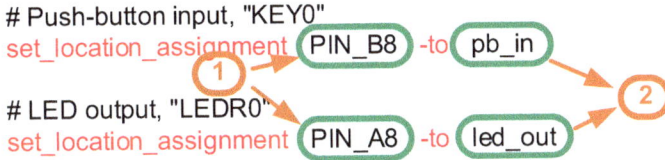

```
# Push-button input, "KEY0"
set_location_assignment PIN_B8 -to pb_in

# LED output, "LEDR0"
set_location_assignment PIN_A8 -to led_out
```

1) these are the pin numbers of the FPGA, "B8" and "A8" (which we get from the board user manual);

2) these are the signal names of our design, and must match (exactly) the signal names as shown in the entity declaration.

The "set_location_assignment" and "-to" entries are keywords for pin assignments.

Launch Quartus, and select "File"/"New Project Wizard":

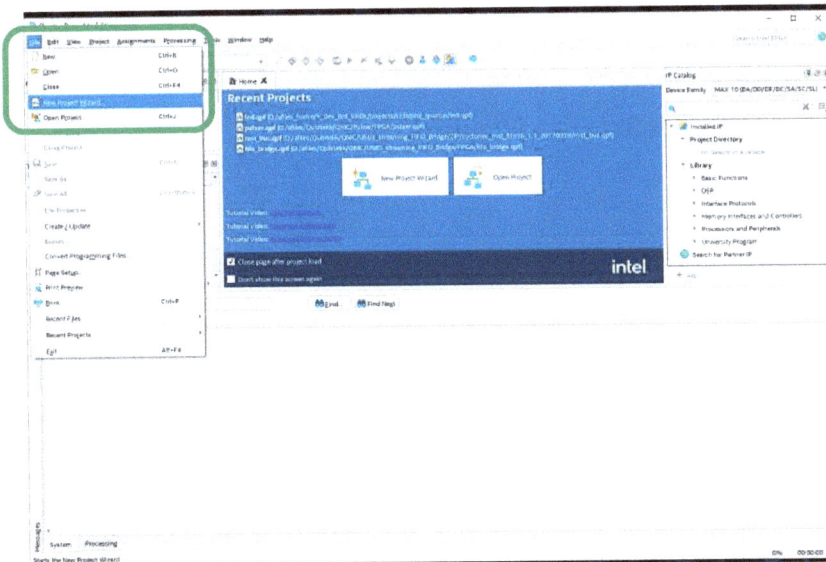

Skip the next introductory window, and enter the project information:

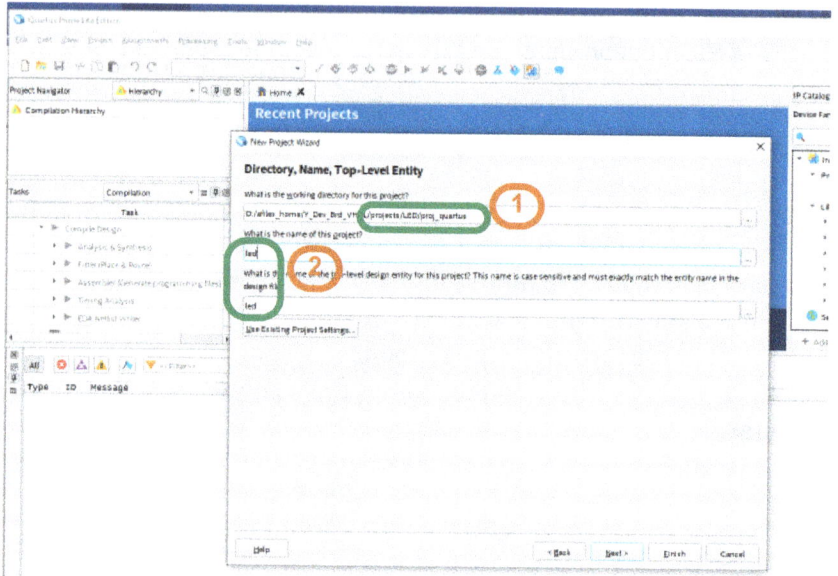

1) this is the project folder we set up initially for Quartus;
2) we normally name our project the same as the highest level VHDL file (in this case, the only source file).

Next, select the default "Empty Project".

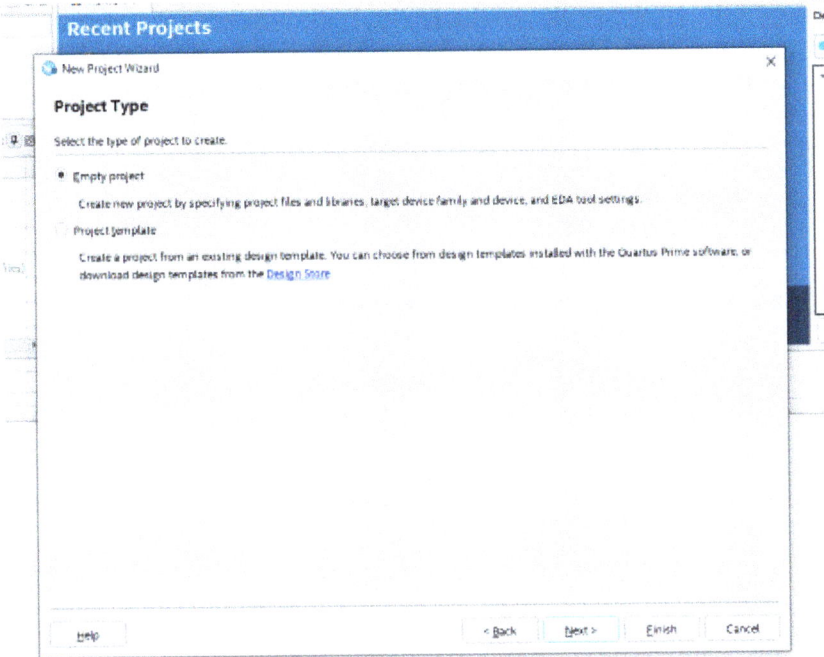

Here we then add our VHDL source files (in this case, just one).

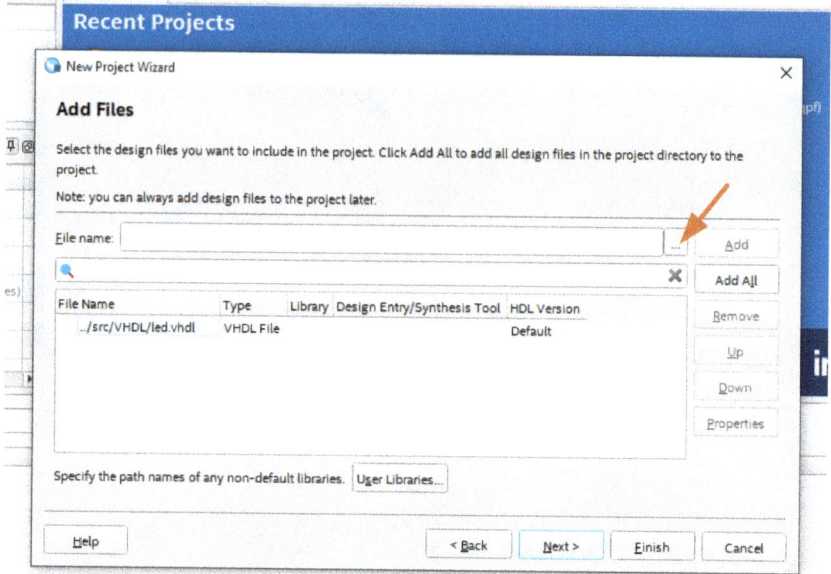

Next we select our specific FPGA device. The part number is included in the Terasic user manual. From the "Family" window, we chose MAX 10 DA.

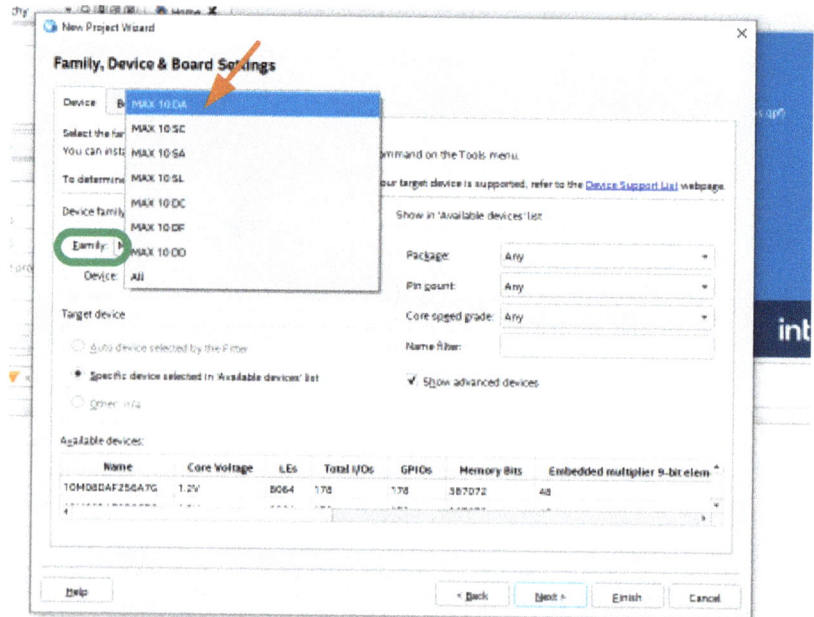

From the user manual, we find that the part number, and under "Available devices," we find and select the closest one: "10M50DAF484C7G".

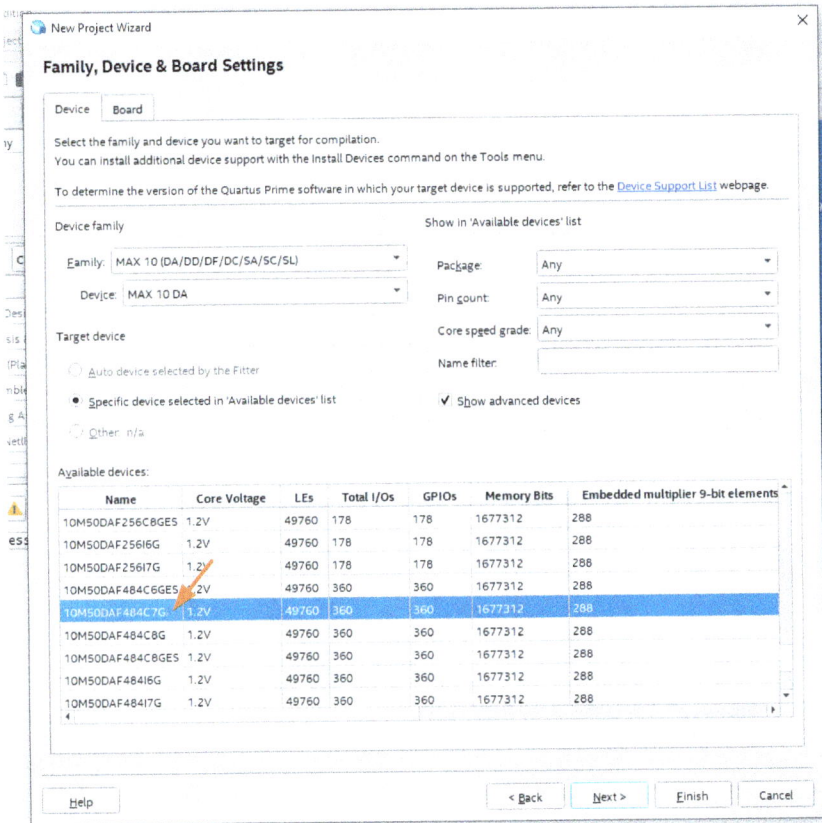

In the next window we select "VHDL".

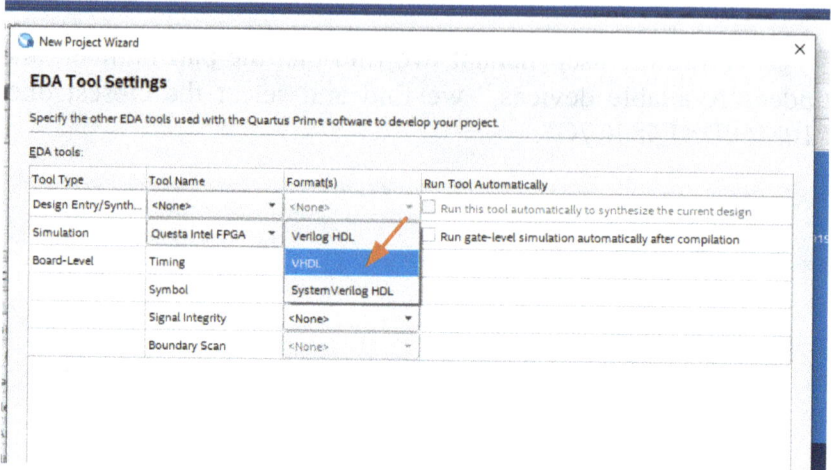

The last window is a summary, and we select "Finish."

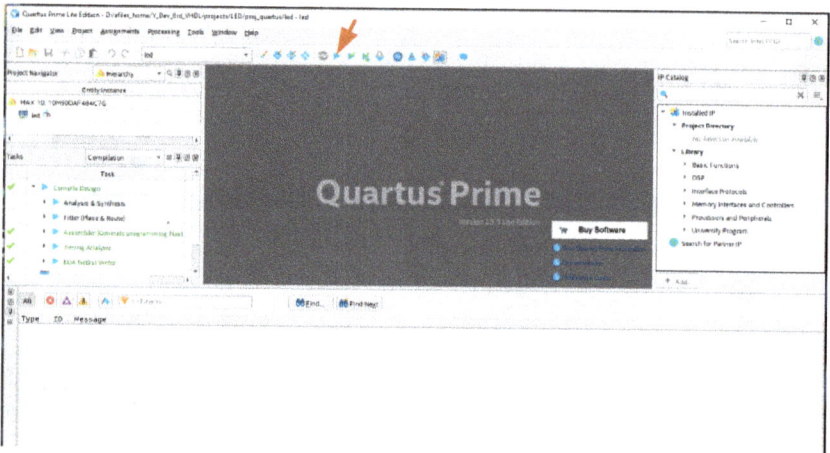

We are now ready to perform a first compile. After compilation, your folder structure should look like this.

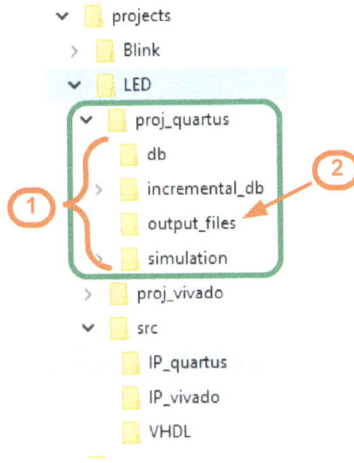

1) Quartus creates these sub-folders;

2) and places the final "led.sof" file in the "output_files" folder. This is the file that we load (configure) onto the FPGA with the Programmer tool.

The Terasic user manual describes the Quartus Programmer tool steps to configure the FPGA using JTAG.

If you've been successful in these somewhat tedious series of tool steps, the board should operate like this.

1) pushing this button,
2) should light this LED.

A small step for the Terasic board, but a giant leap in bringing up the development system.

The Digilent/Vivado project

You have presumably installed Vivado, and are now ready to create and compile a project for our first simple design. There's one additional step before we do, however. We have to tell Vivado what FPGA pins we want to use for the external signals. We do this with what we call a "constraints" file. We must create a text file with a ".xcf" extension, for example, maybe "led.xcf" (although it can be any name, as long as it has the correct extension).

The file you create will contain these lines, which define the FPGA pins:

```
## Buttons
# BTNL
set_property -dict { PACKAGE_PIN W19 IOSTANDARD LVCMOS33 } [get_ports { pb_in }];

## LEDs
# LD0
set_property -dict { PACKAGE_PIN U16 IOSTANDARD LVCMOS33 } [get_ports { led_out }];
```

Everything in red is required.

1) these are the FPGA pins numbers, which we can find in the Digilent user manual;
2) and these are the signals defined in our project entity declaration – they must match exactly.

We'll place this .xcf file at the top of our project hierarchy, under "proj_vivado," although it could go anywhere (we will tell Vivado where it is).

Launch Vivado, and select "Create Project."

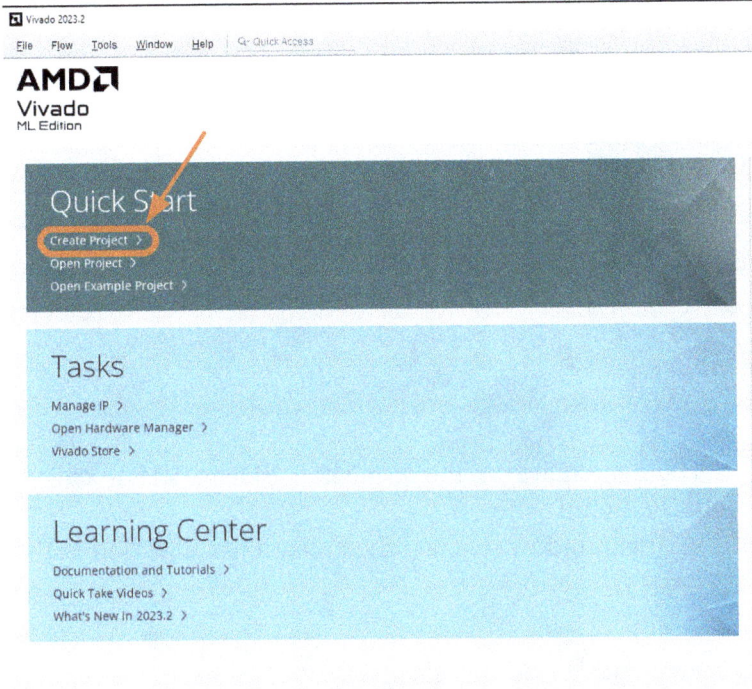

Skip through the next window, and then enter the project name (in this case, we'll just use "led") and folder location.

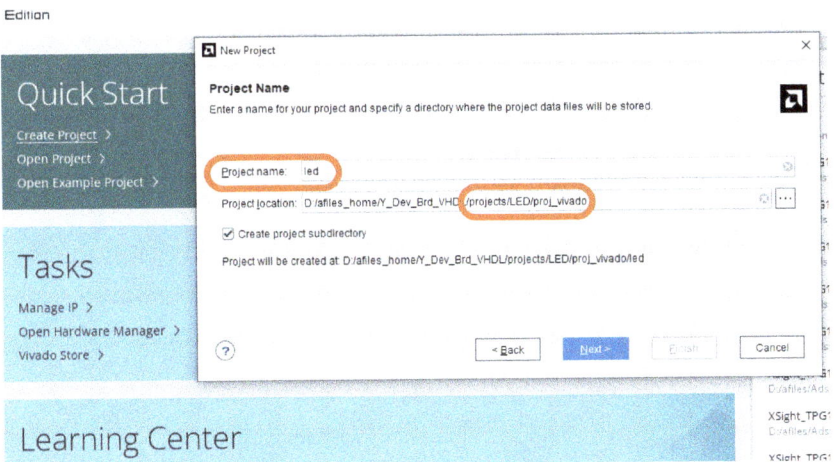

Blaine C. Readler

Choose the default in the next window (RTL Project), and then "Add Files" in the next one.

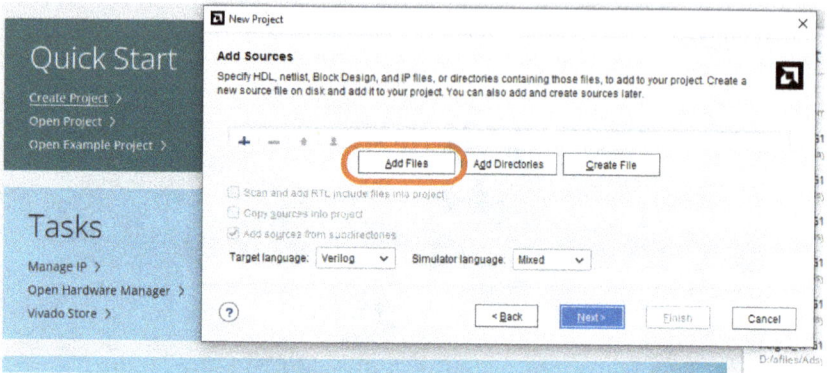

Then navigate to the projects/LED/src/VHDL folder to select the "led.vhdl" file.

In next window, we add the constraints file, "led.xcf".

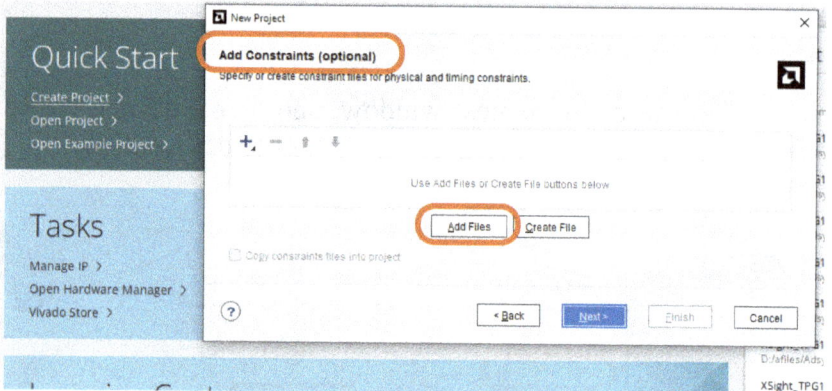

Next, we choose the part that's populated on the Basys-3 board, and we know from the user manual that the family is the Artix-7.

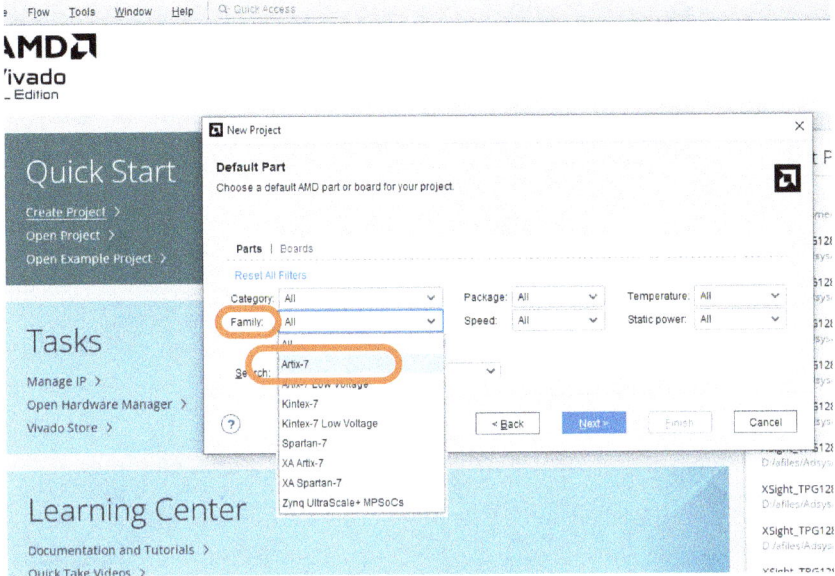

Also from the user manual, we see that the part is indicated to be "XC7A35T-1CPG236C", and from the part list, we find an entry that's the closest.

The last window is a summary, and we finish, ready to compile in the resulting Vivado dashboard window.

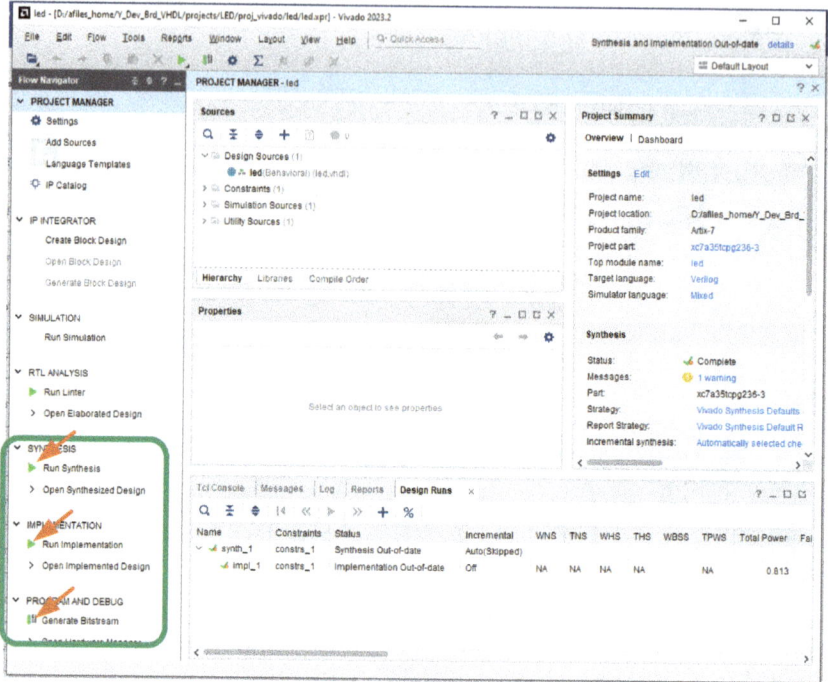

Vivado separates the compilation into "synthesis" (where the tool translates the VHDL code into something that can be laid into the FPGA gates) and "implementation" (the actual layout and routing) steps, and we run them in sequence after each is done. Generating the bitstream is the final step, which creates the final ".bit" file that is loaded into the FPGA.

After compilation, your folder structure should look like this.

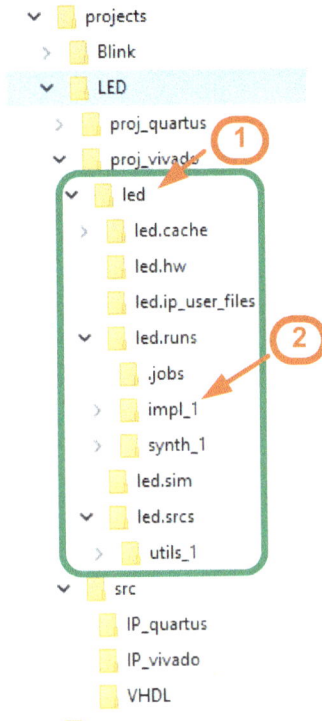

1) Vivado creates the project "led" folder along with the myriad of sub-folders;

2) and places the final "led.bit" file in the implementation folder. This is the file that we load (configure) onto the FPGA with the programmer tool.

The final step is to configure the Digilent FPGA with our compiled .bit file.

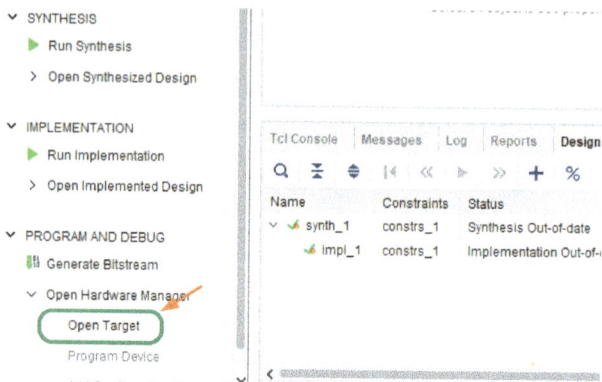

Under "PROGRAM AND DEBUG" select "Open Target"

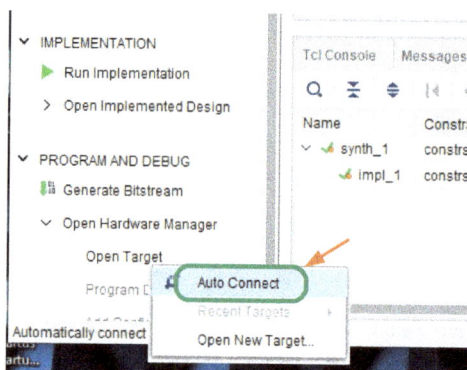

Then "Auto Connect".

At this point, Vivado attempts to connect with the Digilent board's FPGA.

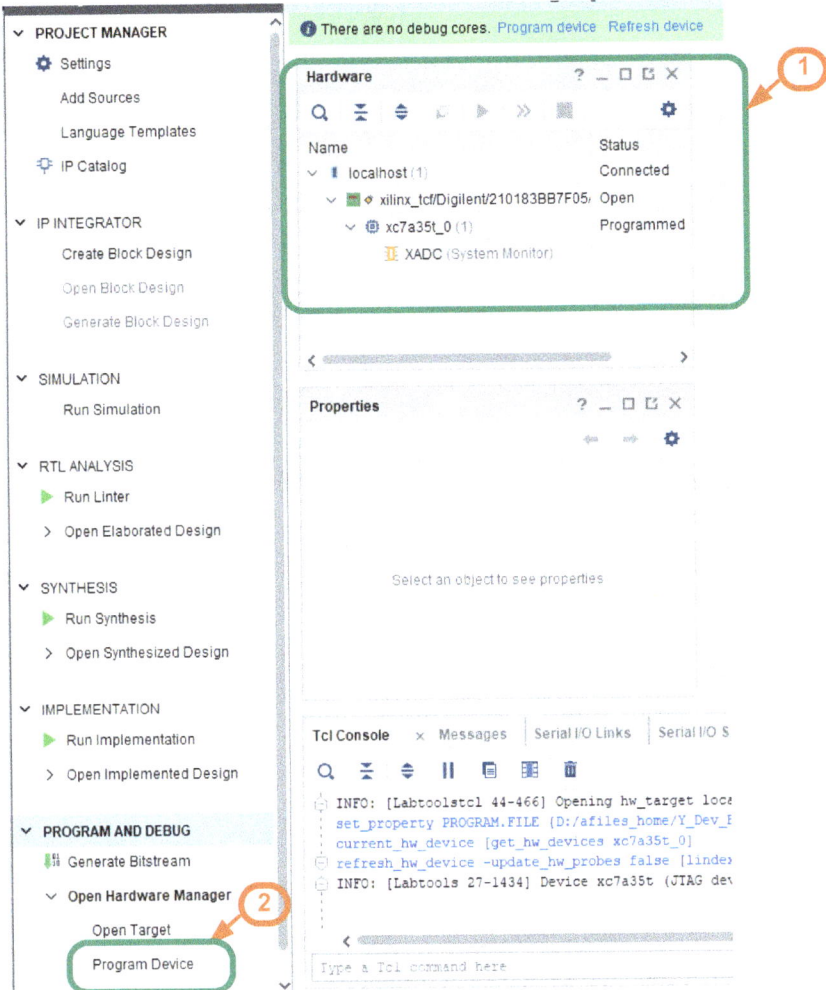

1) if successful, Vivado displays some basic information,
2) and we find "Program Device" is available.

If Vivado wasn't successful in finding the FPGA, check that the USB cable is connected to the board, and that the board is thusly powered. You can also check that "Digilent USB Device" appears in your computer's settings where you find other USB device.
Selecting "Program Device" should result in this.

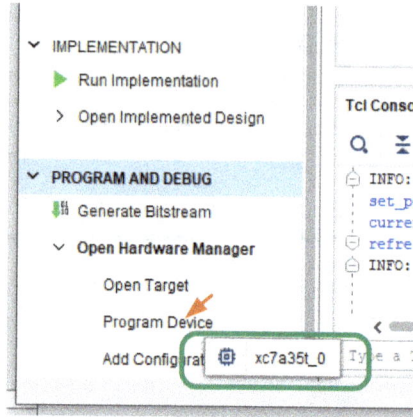

Select the device displayed (e.g., xc7a35t_0, here).

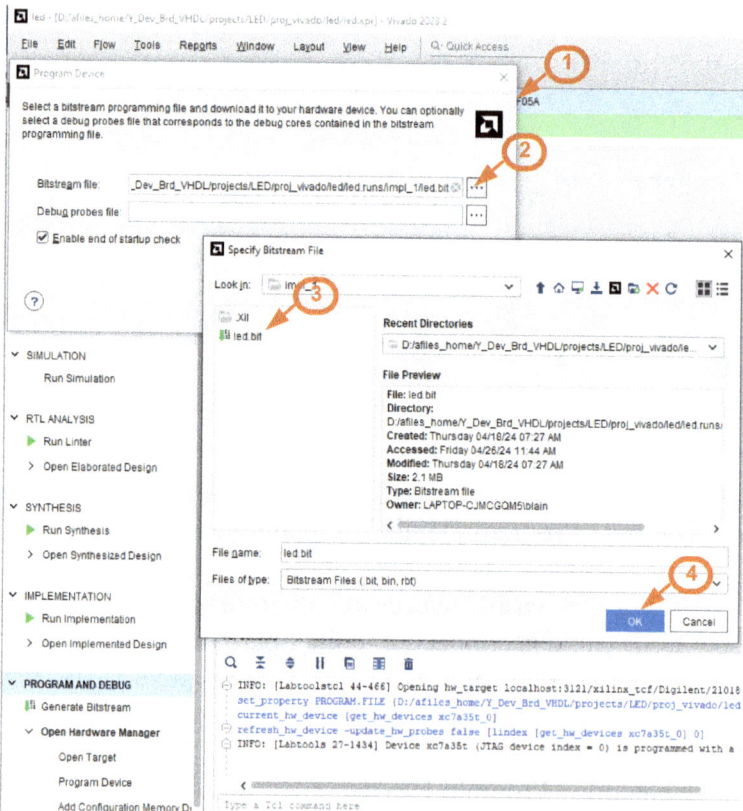

1) in this window,

2) select to find the bitstream file;

3) the program should find the "led.bit" file;

4) select "OK".

Finally, select "Program".

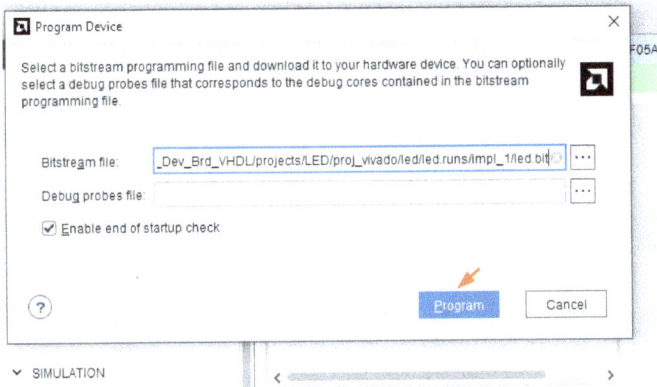

The board's bit file should now download.

If the programming was successful, the board should operate like this.

1) pushing this button,

2) should turn off this LED.

A small step for the Digilent board, but a giant leap in bringing up the development system.

LED Exercises

Try putting logic gating through some paces (e.g. with AND, OR, or XOR) operations between push buttons and slide switches to light different LEDs. You can find one LED exercise example zip file on my website at http://www.readler.com – Engineering / Technical References. Instead of creating a whole new compiler project, you could simply place this (or your own) exercise file in the "src" folder and rename it as "led.vhdl" (remembering to save off the original). Note that my example is just that – yours may be different, and work just fine.

Note that in order to introduce new buttons, switches, and LEDs in your exercise project, you'll also need to add these to either the "*.qsf" (Terasic) or "*.xcf" (Digilent) pin constraint file. These are already included in the exercise project zip file.

Chapter Three
Blink

Our second project steps ahead to introduce a clock and registers. Initially all we'll do is blink an LED. However, like the previous LED project, getting to that first blink sets the stage for much more to follow.

Here's our logic to blink an LED.

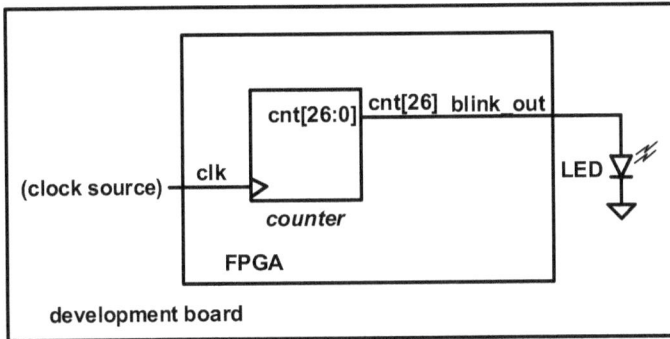

As you know from the free Youtube class, *U of Blaine presents DIGITAL DESIGN*, the counter is composed of twenty-seven registers connected by logic that implements a counting process. Let's take just the first stage and see how we would implement that with VHDL.

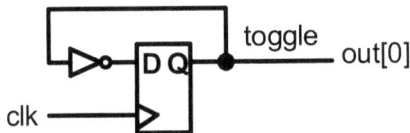

It simply changes polarity, i.e., toggles, with each clock, like so.

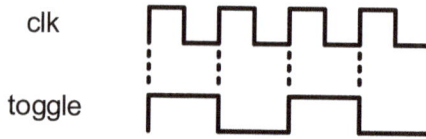

Here's VHDL code that implements this.

```
library IEEE;
use IEEE.STD_LOGIC_1164.all;
use IEEE.STD_LOGIC_MISC.all;

entity blink is
  port
    (
    clk        : in    std_logic;
    --
    blink_out : out   std_logic;
    -- unused LEDs
    led        : out   std_logic_vector(9 downto 1)
    );
end entity;

architecture Behavioral of blink is

    -- signal declarations
    signal toggle : std_logic;

begin

    toggle_process : process(clk)
    begin
      if rising_edge(clk) then
          toggle <= NOT toggle;
      end if;
    end process;

end architecture Behavioral;
```

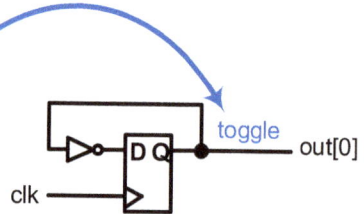

We must first declare the internal toggle signal.

The word "signal" and the colon and ending semi-colon are part of every signal declaration.

1) "toggle" is the signal name;
2) we've already seen "std_logic" declared in the entity declaration.

Now we get to the meat, the clocked toggle register.

```
begin                    process statement
    toggle_process : process (clk)
    begin
        if rising_edge(clk) then
            toggle <= NOT toggle;
        end if;
    end process;
```

Clocked registers are implemented in VHDL process statements, which always include a colon and the words "process," "begin," and "end process;". Additionally, process statements include a sensitivity list, bounded by parenthesis.

```
begin                    process statement
    toggle_process : process (clk)
    begin
        if rising_edge(clk) then        sensitivity list
            toggle <= NOT toggle;
        end if;
    end process;
```

The sensitivity list tells the process when to go "active." In the case of registers, this only occurs at clock edges, so we only need to include the "clk" signal input. Within the process statement, our register is built with a standard IF/THEN conditional statement.

Blaine C. Readler

The IF/THEN structure should, hopefully, be familiar to you, as this is common to all coding languages, from 1958's FORTRAN to modern C++. In this case the "condition" is the rising edge of the clock.

The "rising_edge" with the parenthesis that follow is a keyword that indicates that this process statement only activates at precisely the instant of the clock's rising edge. This, of course, is the definition of a clocked register. You can see how this works.

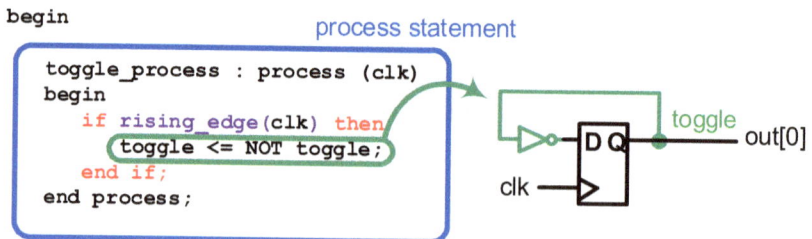

Now that we see how to implement a register in VHDL, we'll add more in order to blink the LED. We'll create a counter that cycles through the blinking time. The clock provided on each development board – 50MHz for the Terasic, and 100MHz for the Digilent – is millions of times faster than what a human can see blinking. We'll use a counter to divide the clock down to a human timescale.

With a 27-stage counter, we can count up to:

0x7FF_FFFF = 134,217,727

At 50MHz (the Terasic board), the LED will blink every 2.7 seconds. At 100MHz (the Digilent board), the LED will blink every 1.35 seconds. We use the MS bit of the counter, which toggles on and off at this rate.

We could create a VHDL counter built up with twenty-seven individual registers, where each stage toggles when all the previous stages are one. A four-stage counter would look like this.

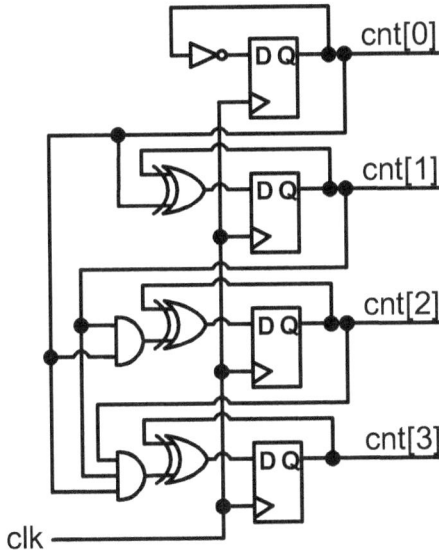

And the corresponding VHDL code might look like this.

Blaine C. Readler

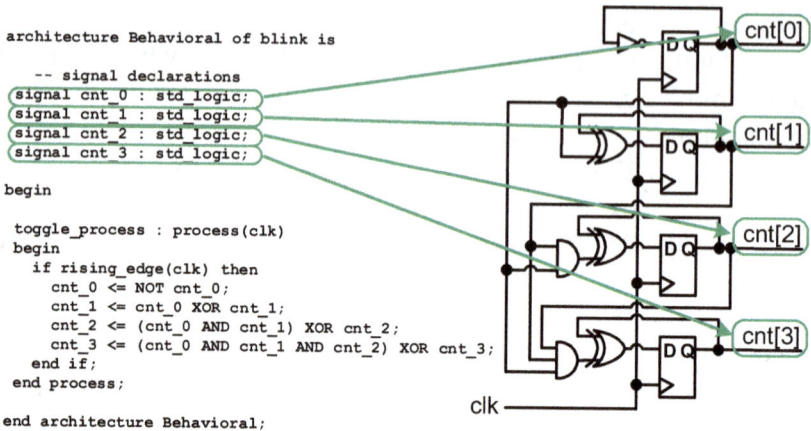

```
architecture Behavioral of blink is

    -- signal declarations
    signal cnt_0 : std_logic;
    signal cnt_1 : std_logic;
    signal cnt_2 : std_logic;
    signal cnt_3 : std_logic;

begin

    toggle_process : process(clk)
    begin
        if rising_edge(clk) then
            cnt_0 <= NOT cnt_0;
            cnt_1 <= cnt_0 XOR cnt_1;
            cnt_2 <= (cnt_0 AND cnt_1) XOR cnt_2;
            cnt_3 <= (cnt_0 AND cnt_1 AND cnt_2) XOR cnt_3;
        end if;
    end process;

end architecture Behavioral;
```

Each signal declaration corresponds to one of the register outputs. And the logic feeding each register stage is like so.

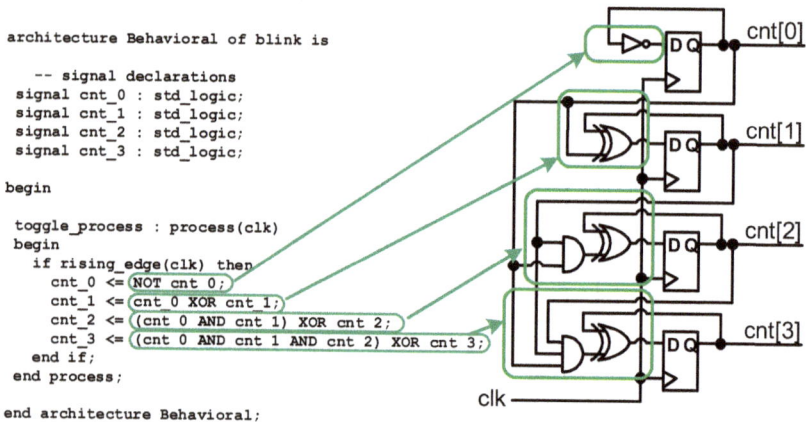

```
architecture Behavioral of blink is

    -- signal declarations
    signal cnt_0 : std_logic;
    signal cnt_1 : std_logic;
    signal cnt_2 : std_logic;
    signal cnt_3 : std_logic;

begin

    toggle_process : process(clk)
    begin
        if rising_edge(clk) then
            cnt_0 <= NOT cnt_0;
            cnt_1 <= cnt_0 XOR cnt_1;
            cnt_2 <= (cnt_0 AND cnt_1) XOR cnt_2;
            cnt_3 <= (cnt_0 AND cnt_1 AND cnt_2) XOR cnt_3;
        end if;
    end process;

end architecture Behavioral;
```

This is cumbersome enough, but imagine what the register logic would like on the last of twenty-seven stages.

42

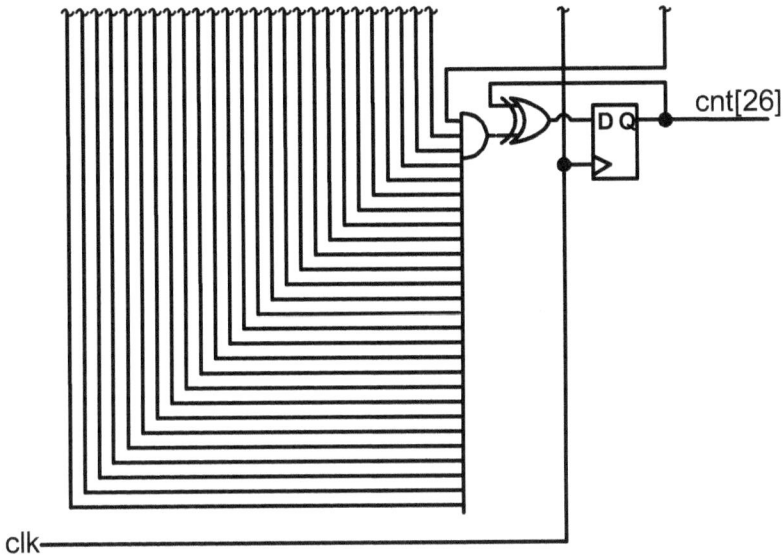

With this as the VHDL statement line.

```
cnt_26 <= (    cnt_0
          AND cnt_1
          AND cnt_2
          AND cnt_3
          AND cnt_4
          AND cnt_5
          AND cnt_6
          AND cnt_7
          AND cnt_8
          AND cnt_9
          AND cnt_10
          AND cnt_11
          AND cnt_12
          AND cnt_13
          AND cnt_14
          AND cnt_15
          AND cnt_16
          AND cnt_17
          AND cnt_18
          AND cnt_19
          AND cnt_20
          AND cnt_21
          AND cnt_22
          AND cnt_23
          AND cnt_24
          AND cnt_25
          ) XOR cnt_26;
```

The entire counter would require about ten pages of code. You might be thinking that there must be a better way, and there is. Simple arithmetic can be implemented in FPGAs relatively easily – addition, subtraction, multiplication (but not so much division) – and VHDL reflects this, providing arithmetic operators. Here's the compact code using simple addition.

```
library IEEE;
use IEEE.STD_LOGIC_1164.all;                    ③
use IEEE.STD_LOGIC_MISC.all;
use IEEE.STD_LOGIC_UNSIGNED.all;
use IEEE.NUMERIC_STD.all;

entity blink is
  port
    (
    clk        : in    std_logic;
    --
    blink_out : out   std_logic;
    -- unused LEDs
    led        : out   std_logic_vector(9 downto 1)
    );
end entity;

architecture Behavioral of blink is              ②

    -- signal declarations
    signal cnt : unsigned(26 downto 0);

begin

    toggle_process : process(clk)
    begin
      if rising_edge(clk) then
        cnt <= cnt + 1;
      end if;                      ①
    end process;

    blink_out <= cnt(26);          ④

end architecture Behavioral;
```

1) each clock period, we increment the count signal by one. This is more or less the definition of a simple counter;

2) here we see "(26 downto 0)" – this is VHDL's cumbersome way of indicating a 27 bit signal. In coding, we call

multi-bit signals "vectors," and single-bit signals "scalars". Normally, we might write this as "[26:0]". Also, instead of a "std_logic" signal, we have "unsigned" – this is a special type of signal that VHDL knows how to do arithmetic on. However, in order to be able to do that arithmetic, we need . . .

3) special libraries. These two, as the names imply, define how to use unsigned numbers, and how to do numeric things like arithmetic. Again, all vendor compiler tools come with these pre-loaded;

4) we use the MS bit of the counter for the output, which toggles on and off at the rate calculated above.

Terasic/Quartus

This LED should be blinking.

Digilent/Vivado

This LED should be blinking.

Blink Exercises

Create blinks of different rates and also blinks with a duty cycle (the ratio of "on" vs "off" times) other than 50/50 by logically combining different bits of the "cnt" vector signal.

As with the LED exercises, you can find one such Blink exercise zip file on my website at http://www.readler.com – Engineering / Technical References.

Also, as with the LED exercises, your design may be different than my example, but if it works, then it's correct.

Chapter Four
Seven-Segment Action

Here, we use two of the seven-segment displays mounted on the boards. When either of two push-buttons is pressed, a corresponding display lights and increments. Thus, when the first button is pressed, the first display becomes active and increments from, say, 2 to 3. If pressed again, then that display increments from 3 to 4. If the second button is pressed, the first display goes inactive, and the second display lights and increments in a similar fashion.

Here's a block diagram.

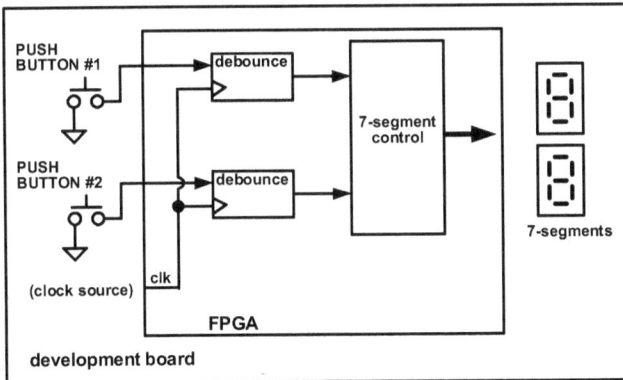

In order to operate smoothly with just one display increment occurring with each button push, we need to filter the button signal with what we call de-bounce logic. The reason is that, although it seems to us that a contact is made directly and securely when we press the button, the contacts actually take a little time to settle together. In other words they bounce a bit. The settling time is very short – thousands of a second – but our logic clock is working even faster, and "sees" each bounce on-and-off.

This next diagram shows what happens.

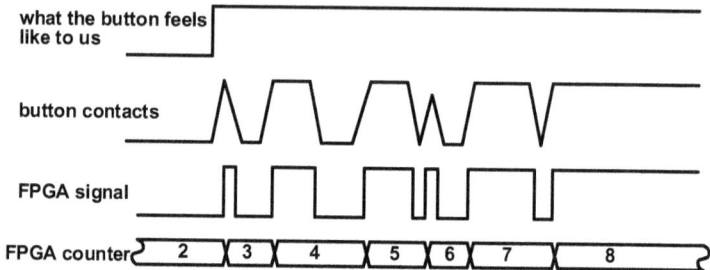

Without debounce filtering, the display quickly increments, and to us – the user – the count seems to jump from two to eight instantaneously.

The next diagram shows what happens when we hold off further actions in the FPGA for a period of time, allowing the button contacts to settle.

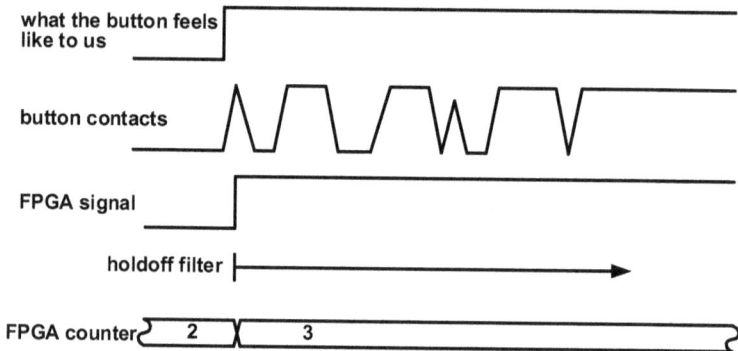

This is code that implements the debounce filter. As we'll see, we kick off a holdoff filter counter with the first button contact, and we then wait until the counter finishes to sample the button, which should have settled by then.

```
debounce_process : process(clk)
begin
```

```
   if rising_edge(clk) then
      pb_in_d1 <= pb_in;
      pb_in_d2 <= pb_in_d1;
      --
      if (pb_in_d2 /= pb_in_d1) then
         hold_cnt <= X"00000";
      elsif (hold_cnt /= X"FFFFF") then
         hold_cnt <= hold_cnt + 1;
      end if;
      --
      if (hold_cnt = X"FFFFE") then
         pb_debounced <= pb_in_d1;
      end if;
   end if;
end process;
```

Let's see how it works.

```
debounce_process : process(clk)
begin
    if rising_edge(clk) then
        pb_in_d1 <= pb_in;          1
        pb_in_d2 <= pb_in_d1;       2
        --
        if (pb_in_d2 /= pb_in_d1) then   3
            hold_cnt <= X"00000";        4
        elsif (hold_cnt /= X"FFFFF") then
            hold_cnt <= hold_cnt + 1;    5
        end if;
        --
        if (hold_cnt = X"FFFFE") then
            pb_debounced <= pb_in_d1;
        end if;
    end if;
end process;
```

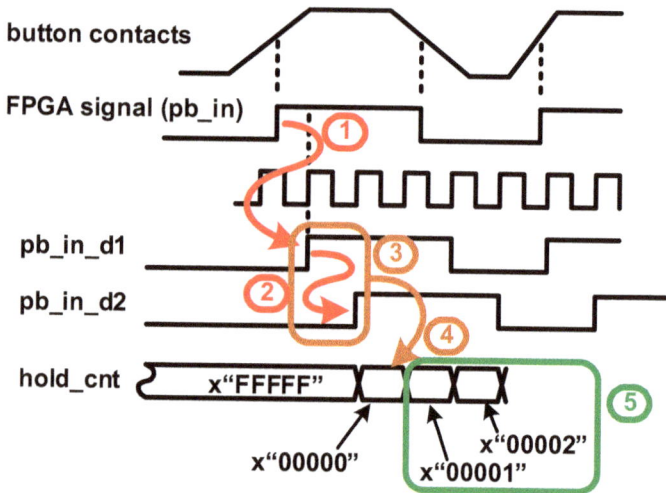

1) the input signal to the FPGA is "pb_in," and we clock it once as "pb_in_d1" to synchronize it to our clock – a standard step for asynchronous inputs;

2) we clock "pb_in_d1" once again into another register as "pb_in_d2", and you can see that as the button goes high for the first time, "pb_in_d1" is high while "pb_in_d2" is still low for one clock until "pb_in_d2" also goes high;

3) we use that condition,

4) to clear our holdoff filter counter,

5) which then begins counting up.

Eventually we come to the last button contact bounce.

1) the counter has reached 0x150, but is cleared again (as it has been with each bounce);

2) and then is cleared one last time as the button input finally becomes stable.

Finally, the counter manages to complete its count.

```
debounce_process : process(clk)
begin
  if rising_edge(clk) then
    pb_in_d1 <= pb_in;
    pb_in_d2 <= pb_in_d1;
    --
    if (pb_in_d2 /= pb_in_d1) then
       hold_cnt <= X"00000";
    elsif (hold_cnt /= X"FFFFF") then
       hold_cnt <= hold_cnt + 1;           1
    end if;
    --
    if (hold_cnt = X"FFFFE") then           2
       pb_debounced <= pb_in_d1;
    end if;                                  3
  end if;
end process;
```

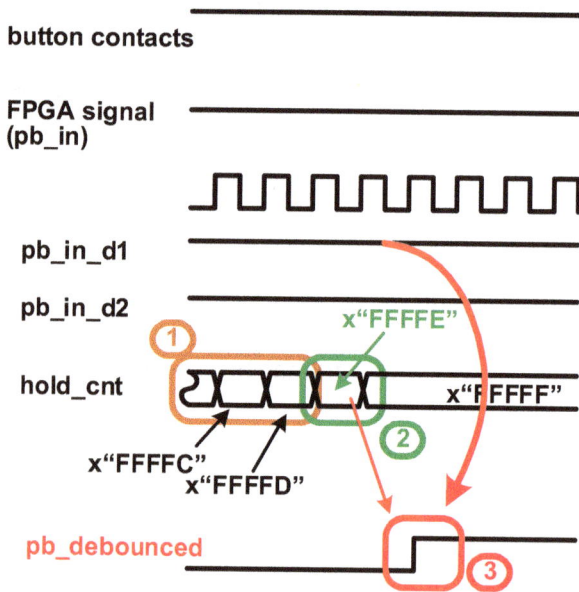

1) the last counts are on their way to the all-Fs terminal value;

2) we detect that the count is one from the end (0xFFFFF – 1),

3) and finally sample the button input to set the debounced output, "pb_debounced."

Note that this code works to debounce the push-button both when it's pushed and also when it's released. Otherwise the counter might make a flurry of counts when the button is lifted.

We note that the Basys board includes a board-based debounce circuit for its push-buttons (analog filter and a Schmidt trigger buffer for clean logic level detection), so in this case, our debounce circuit is supplementary (and still instructional).

Next we'll look at the code for the 7-segment's project, and we introduce the concept of hierarchical design using module components.

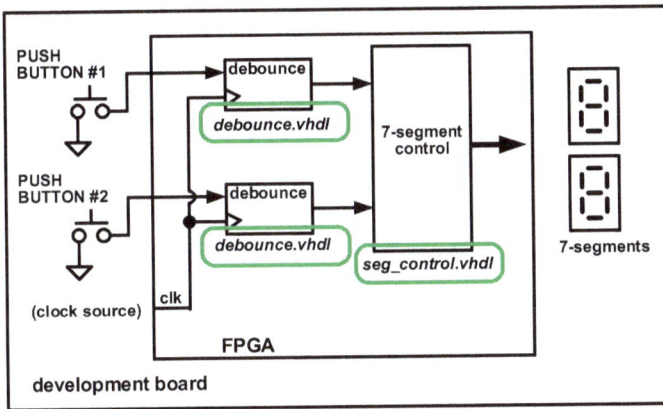

As we see here, our components correspond to the sub-blocks in the diagram above: debounce (debounce.vhdl) and 7-segment control (seg_control.vhdl). A hierarchical approach provides improved comprehension of operation – subsuming the clutter of details into clear functional blocks. Besides this, though, hierarchy provides other benefits. For example, for logic that is used more than once, as is the case here with the debounce block.

We've reached the point where we can't always see an entire code file at one time – some just don't fit on the book's page. And so, we'll look at the various VHDL file structure sections as originally described:

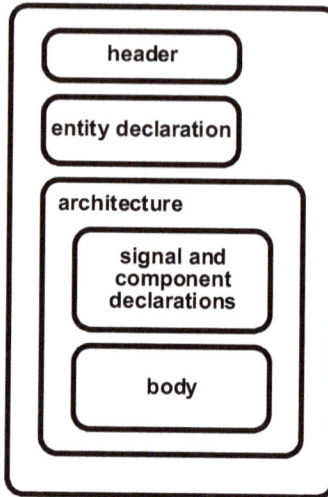

Here's the entity declaration for our top-level file, called "seven_segment."

entity declaration

```
entity seven_segment is
  port
    (
    clk              : in    std_logic;
    pb_1             : in    std_logic;
    pb_2   (1)       : in    std_logic;
    --
    seven_seg_1  (2.) out   std_logic_vector(7 downto 0);
    seven_seg_2      out   std_logic_vector(7 downto 0);
    en1              : out   std_logic;
    en2    (3)       : out   std_logic;
    en3              : out   std_logic;
    en4              : out   std_logic;
    -- unused LEDs
    led          : out   std_logic_vector(9 downto 0)
    );
end entity;
```

1) our two push-button inputs;
2) the 7-segment outputs, and;

3) enables for the seven-segment displays (used only on the Digilent board).

We see a new signal type for the 7-segment outputs, "std_logic_vector". This simply indicates that the signal is a vector, and the bit field indications ("7 downto 0") have the same meaning as we saw with unsigned types.

Here's the signal and component declarations.

signal and component declarations

```
architecture Behavioral of seven_segment is

  -- component declarations

  component debounce is
  port                            (2)
   (
    clk          : in    std_logic;
    pb_in        : in    std_logic;
    --
    pb_debounced : out   std_logic
   );
  end component;

  component seg_control is
  port                            (3)
   (
    clk          : in    std_logic;
    pb_debncd_1  : in    std_logic;
    pb_debncd_2  : in    std_logic;
    --
    seven_seg_1  : out   std_logic_vector(7 downto 0);
    seven_seg_2  : out   std_logic_vector(7 downto 0);
    en1          : out   std_logic;
    en2          : out   std_logic
   );
  end component;

  -- signal declarations
  signal pb_debncd_1 : std_logic;   (1)
  signal pb_debncd_2 : std_logic;
```

1) these are the signals at the top level that connect the debounce modules to the 7-segment control modules;

57

2) the debounce component. This is the "debounce" box in the diagram above. We must declare it here in order to use it in the body that follows. Note that the structure looks very much like the entity declaration for the top file above. The component associated with this declaration is actually just another whole VHDL source file. VHDL files can stand alone (as the LED and Blink did), or can be instantiated within higher-level files (as this one is), and this component declaration is indeed essentially the entity declaration of that "debounce.vhdl" file (which we'll get to eventually). When we use it as a component in a higher-level file, we simply replace "entity" with "component";

3) the 7-segment control module.

Lastly, we look at the VHDL body.

architecture body

```
begin

    debounce_1 : debounce
    port map
      (
      clk              => clk,          --in    std_logic;
      pb_in            => pb_1,         --in    std_logic;
      --
      pb_debounced => pb_debncd_1   --out   std_logic
      ) ;

    debounce_2 : debounce
    port map
      (
      clk              => clk,          --in    std_logic;
      pb_in            => pb_2,         --in    std_logic;
      --
      pb_debounced => pb_debncd_2   --out   std_logic
      ) ;

    seg_control_i : seg_control
    port map
      (
      clk            => clk,          --in std_logic;
      pb_debncd_1    => pb_debncd_1,  --in std_logic;
      pb_debncd_2    => pb_debncd_2,  --in std_logic;
      --
      seven_seg_1  => seven_seg_1,  --out std_logic_vector(7 downto 0);
      seven_seg_2  => seven_seg_2,  --out std_logic_vector(7 downto 0);
      en1          => en1,          --out std_logic;
      en2          => en2           --out std_logic
      ) ;

end architecture Behavioral;
```

We find that it is nothing more than the declared components stitched together. We'll take the first debounce component and see how we connect the various signals to it.

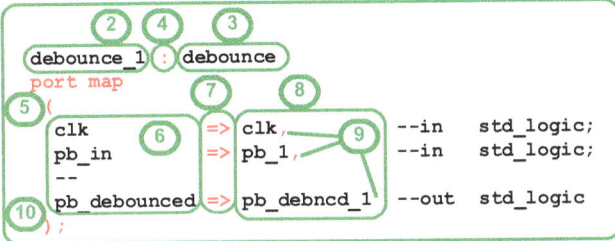

```
begin

  debounce_1 : debounce
  port map
  (
    clk          => clk,        --in   std_logic;
    pb_in        => pb_1,       --in   std_logic;
    --
    pb_debounced => pb_debncd_1 --out  std_logic
  );

  debounce_2 : debounce
  port map
  (
    clk          => clk,        --in   std_logic;
    pb_in        => pb_2,       --in   std_logic;
    --
    pb_debounced => pb_debncd_2 --out  std_logic
  );

  seg_control_i : seg_control
  port map
  (
    clk          => clk,            --in std_logic;
    pb_debncd_1  => pb_debncd_1,    --in std_logic;
    pb_debncd_2  => pb_debncd_2,    --in std_logic;
    --
    seven_seg_1  => seven_seg_1,    --out std_logic_vector(7 downto 0);
    seven_seg_2  => seven_seg_2,    --out std_logic_vector(7 downto 0);
    en1          => en1,            --out std_logic;
    en2          => en2             --out std_logic
  );

end architecture Behavioral;
```

1) the first of two instantiated debounce components;

2) this is the name of this particular instantiation, "debounce_1". This name can be anything, but must be unique to this instantiation. You'll find it useful, however, to choose a name that somehow indicates the instantiated component, since development and simulation tools often will refer to the instantiation name in their messaging;

3) this is the name of the actual component, i.e., this is the entity name in the component's own VHDL file;

59

4) a colon separates the two, and is a keyword;

5) "port map" and the "(" parenthesis are keywords, flagging that signal connections follow;

6) these are the signals from above in the component declaration;

7) the "=>" combination are keywords, and indicate the signal connections between the component and signals local to this file;

8) the local signals, either from those declared above in the declaration section (such as "pb_debncd_1"), or input/output signals of this file (signals shown in the entity declaration above, such as "clk" and "pb_1");

9) the local signals are followed by a comma, all except the last one;

10) the end of the component instantiation is marked by a closing parenthesis and semi-colon.

The following diagram shows the signal connections.

```
begin                         (entity declaration

  debounce_1 : debounce
  port map
    (
    clk            => clk,              --in    std_logic;
    pb_in          => pb_1,             --in    std_logic;
    --
    pb_debounced  => pb_debncd_1    --out   std_logic
    ) ;

  debounce_2 : debounce
  port map
    (
    clk            => clk,              --in    std_logic;
    pb_in          => pb_2,             --in    std_logic;
    --
    pb_debounced  => pb_debncd_2    --out   std_logic
    ) ;

  seg_control_i : seg_control
  port map
    (
    clk            => clk,              --in std_logic;
    pb_debncd_1    => pb_debncd_1,  --in std_logic;
    pb_debncd_2    => pb_debncd_2,  --in std_logic;
    --
    seven_seg_1    => seven_seg_1,  --out std_logic_vector(7 downto 0);
    seven_seg_2    => seven_seg_2,  --out std_logic_vector(7 downto 0);
    en1            => en1,              --out std_logic;
    en2            => en2               --out std_logic
    ) ;

end architecture Behavioral;
```

Next we dive into the debounce module, i.e., the debounce entity. We've already seen an architecture body (at the beginning of the chapter), which consists of just one clocked process. Here's the entity and architecture signals declarations.

61

```
entity debounce is
  port
    (
    clk    (1)        : in    std_logic;
    (pb_in)           : in    std_logic;
    --
    (pb_debounced): out   std_logic
    );        (2)
end entity;

architecture Behavioral of debounce is
                                        (3)
    signal pb_in_d1 : std_logic;
    signal pb_in_d2 : std_logic;
    signal hold_cnt : unsigned(19 downto 0); -- ~10-20ms

begin
```

1) the "pb_in" signal connected to the entity input of the top-level "seven_segment" entity;
2) the debounced output connected to the "seg_control" module;
3) the local signals for this entity/architecture.

We finally come to the "seg_control" module, and here we must pause a moment. Our two boards – Xilinx-based Digilent, and Altera-based Terasic – use different 7-segment display components. Each display component on the Terasic board has its own 8-bit display bus input, while the displays on the Digilent board share one 8-bit bus among them. Thus, we need a different "seg_control" VHDL file for each board. And here we find yet another advantage of hierarchy – by separating the "seg_control" operation out as an instantiated module, our shared overall design can point to one "seg_control" file or the other, depending on which board we're compiling for. This is the reason that we created the separate source IP folders for the two boards in our original folder structure:

We place each "seg_control" VHDL file in its appropriate folder, and "point" to the appropriate file within our compiler tool (either Quartus or Vivado). So, for example, depending on which board you choose to go with, you have no use for the other file and associated IP folder.

A point of clarification, though: we have a folder called "VHDL" that is parallel to the individual IP folders, and, of course, our "seg_control" file is also VHDL. As it happens, these IP folders can hold design file types other than VHDL (e.g., IP created from within the vendor tool).

Moving on, we understand that we have two "seg_control" files – similar, but with differences on the specifics of the 7-segment display control. They share a common core operation, however. Both have two counters that track the two button pushes, and both translate these counts into the proper segments to light in the display. The method we use to perform this translation is, as we see here, a "case" statement, which is essentially a lookup table.

```
case (display_cnt) is          -- 76543210
    when X"0" => seven_seg <= "11000000";
    when X"1" => seven_seg <= "11111001";
    when X"2" => seven_seg <= "10100100";
    when X"3" => seven_seg <= "10110000";
    when X"4" => seven_seg <= "10011001";
    when X"5" => seven_seg <= "10010010";
    when X"6" => seven_seg <= "10000010";
    when X"7" => seven_seg <= "11111000";
    when X"8" => seven_seg <= "10000000";
    when X"9" => seven_seg <= "10010000";
    when X"A" => seven_seg <= "10001000";
    when X"b" => seven_seg <= "10000011";
    when X"C" => seven_seg <= "11000110";
    when X"d" => seven_seg <= "10100001";
    when X"E" => seven_seg <= "10000110";
    when X"F" => seven_seg <= "10001110";
    when others => seven_seg <= "11111111";
end case;
```

Key parts of any case statement are indicated in red. Here's how it works.

```
                1
case (display_cnt) is          -- 76543210
    when X"0" => seven_seg <= "11000000";
    when X"1" => seven_seg <= "11111001";
    when X"2" => seven_seg <= "10100100";    3
    when X"3" => seven_seg <= "10110000";
    when X"4" => seven_seg <= "10011001";
    when X"5" => seven_seg <= "10010010";
  2 when X"6" => seven_seg <= "10000010";
    when X"7" => seven_seg <= "11111000";
    when X"8" => seven_seg <= "10000000";
    when X"9" => seven_seg <= "10010000";
    when X"A" => seven_seg <= "10001000";
    when X"b" => seven_seg <= "10000011";
    when X"C" => seven_seg <= "11000110";
    when X"d" => seven_seg <= "10100001";
    when X"E" => seven_seg <= "10000110";
    when X"F" => seven_seg <= "10001110";
    when others => seven_seg <= "11111111";
end case;
```

1) the signal inside the parenthesis can be thought of as the table lookup address or location value;

2) these are the values that point to specific locations in the lookup table; and

3) these are the contents of the lookup table, and consist of any legitimate VHDL statement – here (and quite often) simply an assignment statement.

Let's take an example.

```
       ①   "0111"                        ③
   case(display_cnt) is      -- (76543210)
      when X"0" => seven_seg <= "11000000";
      when X"1" => seven_seg <= "11111001";
      when X"2" => seven_seg <= "10100100";
      when X"3" => seven_seg <= "10110000";
      when X"4" => seven_seg <= "10011001";
      when X"5" => seven_seg <= "10010010";
      when X"6" => seven_seg <= "10000010";
      when X"7" => (seven_seg <= "11111000";) ③
    ② when X"8" => seven_seg <= "10000000";
      when X"9" => seven_seg <= "10010000";
      when X"A" => seven_seg <= "10001000";
      when X"b" => seven_seg <= "10000011";
      when X"C" => seven_seg <= "11000110";
      when X"d" => seven_seg <= "10100001";
      when X"E" => seven_seg <= "10000110";
      when X"F" => seven_seg <= "10001110";
      when others => seven_seg <= "11111111";
   end case;
```

1) let's say that "display_cnt" happens to be a binary seven;

2) in that case, this entry is selected,

3) and therefore, the "seven_seg" signal is assigned this 8-bit value.

And, what happens to values assigned to "seven_seg"?

"0111"

①

```
case(display_cnt) is            -- 76543210
   when X"0" => seven_seg <= "11000000";
   when X"1" => seven_seg <= "11111001";
   when X"2" => seven_seg <= "10100100";
   when X"3" => seven_seg <= "10110000";
   when X"4" => seven_seg <= "10011001";
   when X"5" => seven_seg <= "10010010";
   when X"6" => seven_seg <= "10000010";
   when X"7" => seven_seg <= "11111000";
   when X"8" => seven_seg <= "10000000";
   when X"9" => seven_seg <= "10010000";
   when X"A" => seven_seg <= "10001000";
   when X"b" => seven_seg <= "10000011";
   when X"C" => seven_seg <= "11000110";
   when X"d" => seven_seg <= "10100001";
   when X"E" => seven_seg <= "10000110";
   when X"F" => seven_seg <= "10001110";
   when others => seven_seg <= "11111111";
end case;
```

②

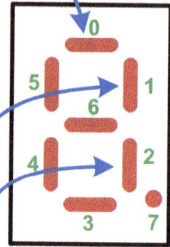

1) in the code, this is just a comment indicating the bit position of the 8-bit assignment values;

2) since the activation of individual LED segments is low-active, for a case value of 0x7 (binary "0111"), the "seven_seg" assignment lights these segments.

And this produces this on the display.

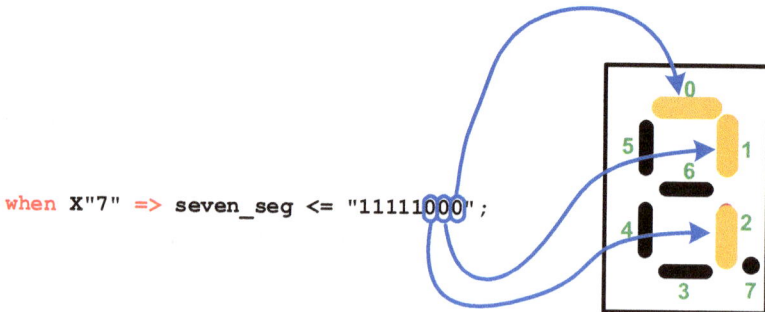

```
when X"7" => seven_seg <= "11111000";
```

Notice that, although the standard coding label indication for hexadecimal values is "0x", in VHDL it is just "X" and the hexadecimal value enclosed by quotation marks. Thus, 0x7 is the same as VHDL's X"7", is the same as VHDL's "0111".

Now that we understand the basic mechanism for translating our push-button incremented counts to displays, let's look at the VHDL code for each of the two boards.

Terasic "seg_control"

Here's how the 7-segment displays are driven on the Terasic board.

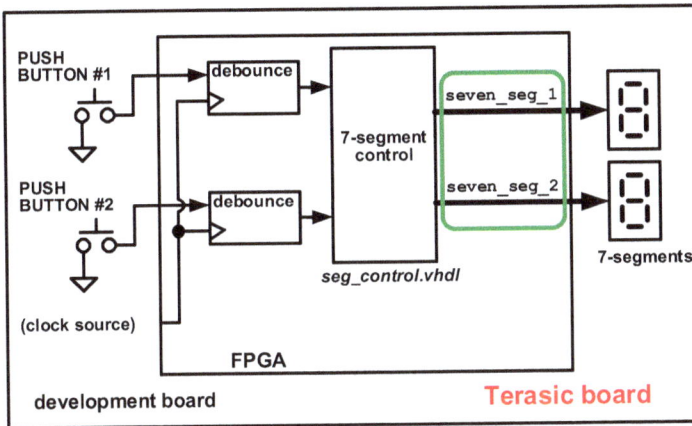

Each 7-segment display is driven by its own dedicated 8-bit bus signal.

Here's the entity declaration:

```
entity seg_control is
  port
    (
    clk           : in   std_logic;
    pb_debncd_1   : in   std_logic;
    pb_debncd_2   : in   std_logic;
    --
    seven_seg_1   : out  std_logic_vector(7 downto 0);
    seven_seg_2   : out  std_logic_vector(7 downto 0);
    en1           : out  std_logic; --not used on this board
    en2           : out  std_logic  --not used on this board
    );
end entity;
```

1) the debounced push-button inputs;

67

Blaine C. Readler

2) the 7-segment buses driving the external displays;

3) these signals are not used on the Terasic board, but since we use the same component instantiation at the top level ("seven_segment.vhdl"), we must include them here.

The architecture signal declarations.

```
architecture Behavioral of seg_control is
    signal pb_debncd_1_d1 :  std_logic;
    signal pb_debncd_2_d1 :  std_logic;
    signal display_select :  std_logic; -- 0 = #1, 1 = #2
    signal display_cnt_1  :  unsigned(3 downto 0);
    signal display_cnt_2  :  unsigned(3 downto 0);
```

1) used to detect a push-button event;

2) this signal tracks which push-button has last been pushed;

3) these are the counts that are displayed.

Here's the architecture body, but don't try to make out the tiny letters – we'll zoom in on the sections:

A) a couple of straight assignments; and

B) a first process to increment the two counts and toggle the select signal, and

C) a second process to implement the case statement display lookup tables that we already saw.

```vhdl
begin

  en1 <= '0'; -- not used on the Terasic board    (A)
  en2 <= '0';

  seg_control_process : process(clk)
  begin
    if rising_edge(clk) then
      pb_debncd_1_d1 <= pb_debncd_1;
      pb_debncd_2_d1 <= pb_debncd_2;
      --
      if    (    pb_debncd_1    = '0' --Terasic PBs are low-active
             AND pb_debncd_1_d1 = '1'
            ) then
            display_select <= '0';
            display_cnt_1  <= display_cnt_1 + 1;
      elsif (    pb_debncd_2    = '0' --Terasic PBs are low-active
             AND pb_debncd_2_d1 = '1'
            ) then
            display_select <= '1';
            display_cnt_2  <= display_cnt_2 + 1;
      end if;
    end if;
  end process;

  --     0
  --   5   1                                        (B)
  --     6
  --   4   2
  --     3

  seg_decode_process : process(clk)              (C)
  begin
    if rising_edge(clk) then
      if (display_select = '1') then
        seven_seg_1 <= "11111111";
      else -- display segments are low-active
        case(display_cnt_1) is        -- 76543210
          when X"0" => seven_seg_1 <= "11000000";
          when X"1" => seven_seg_1 <= "11111001";
          when X"2" => seven_seg_1 <= "10100100";
          when X"3" => seven_seg_1 <= "10110000";
          when X"4" => seven_seg_1 <= "10011001";
          when X"5" => seven_seg_1 <= "10010010";
          when X"6" => seven_seg_1 <= "10000010";
          when X"7" => seven_seg_1 <= "11111000";
          when X"8" => seven_seg_1 <= "10000000";
          when X"9" => seven_seg_1 <= "10010000";
          when X"A" => seven_seg_1 <= "10001000";
          when X"b" => seven_seg_1 <= "10000011";
          when X"C" => seven_seg_1 <= "11000110";
          when X"d" => seven_seg_1 <= "10100001";
          when X"E" => seven_seg_1 <= "10000110";
          when X"F" => seven_seg_1 <= "10001110";
          when others => seven_seg_1 <= "11111111";
        end case;
      end if;
      --
      if (display_select = '0') then
        seven_seg_2 <= "11111111";
      else
        case(display_cnt_2) is        -- 76543210
          when X"0" => seven_seg_2 <= "11000000";
          when X"1" => seven_seg_2 <= "11111001";
          when X"2" => seven_seg_2 <= "10100100";
          when X"3" => seven_seg_2 <= "10110000";
          when X"4" => seven_seg_2 <= "10011001";
          when X"5" => seven_seg_2 <= "10010010";
          when X"6" => seven_seg_2 <= "10000010";
          when X"7" => seven_seg_2 <= "11111000";
          when X"8" => seven_seg_2 <= "10000000";
          when X"9" => seven_seg_2 <= "10010000";
          when X"A" => seven_seg_2 <= "10001000";
          when X"b" => seven_seg_2 <= "10000011";
          when X"C" => seven_seg_2 <= "11000110";
          when X"d" => seven_seg_2 <= "10100001";
          when X"E" => seven_seg_2 <= "10000110";
          when X"F" => seven_seg_2 <= "10001110";
          when others => seven_seg_2 <= "11111111";
        end case;
      end if;
    end if;
  end process;

end architecture Behavioral;
```

Blaine C. Readler

We'll take each of these in turn.

Ⓐ
```
en1 <= '0'; -- not used on the Terasic board
en2 <= '0';
```

As the comment indicates, these signals are not used on the Terasic board, but since we must connect them to *something* on the board (in order to use the same top-level file), I've connected them to a couple of slide switches (via the Quartus "7-segment.qsf" file).

Next we look at the first of the two processes.

Ⓑ
```
seg_control_process : process(clk)
begin
    if rising edge(clk) then
        pb_debncd_1_d1 <= pb_debncd_1;        ①
        pb_debncd_2_d1 <= pb_debncd_2;
        --
        if    (     pb_debncd_1    = '0' --Terasic PBs are low-active
              AND pb_debncd_1_d1 = '1'
              ) then                            ②
            display_select <= '0';
            display_cnt_1   <= display_cnt_1 + 1;
        elsif (     pb_debncd_2    = '0' --Terasic PBs are low-active
              AND pb_debncd_2_d1 = '1'
              ) then
            display_select <= '1';
            display_cnt_2 <= display_cnt_2 + 1;
        end if;
    end if;
end process;
```

1) we develop a one-clock delayed version of the debounced button signals for each of the two button inputs,

2) which we use to create a one-clock event when a button is pushed. We saw something like this back in the debounce module, but whereas there we detected when the inputs button signal simply changed (went either high or low), here we specifically detect that the input debounced signal goes from high to low (the push-buttons on the Terasic board are low-active) – i.e., we find the one clock period where the input has gone low, but its delayed version is still high.

```
seg_control_process : process(clk)
begin
   if rising_edge(clk) then
      pb_debncd_1_d1 <= pb_debncd_1;
      pb_debncd_2_d1 <= pb_debncd_2;
      --
      if    (      pb_debncd_1    = '0' --Terasic PBs are low-active
            AND pb_debncd_1_d1 = '1'
      ) then
         display_select <= '0';
         display_cnt_1   <= display_cnt_1 + 1;
      elsif (      pb_debncd_2    = '0' --Terasic PBs are low-active
            AND pb_debncd_2_d1 = '1'
      ) then
         display_select <= '1';
         display_cnt_2   <= display_cnt_2 + 1;
      end if;
   end if;
end process;
```

At each of these one-clock button-pushed events,

1) we either clear or set the signal that selects which display to light (low for the first display, and high for the second), and

2) increment the counter associated with the button that was pushed.

As we now see, this hexadecimal count value is the source for the ASCII count value displayed.

71

(C)

```
seg_decode_process : process(clk)
begin
    if rising_edge(clk) then
        if (display_select = '1') then          ①
            seven_seg_1 <= "11111111";              ②
        else -- display segments are low-active
            case(display_cnt_1) is          -- 76543210
                when X"0" => seven_seg_1 <= "11000000";
                when X"1" => seven_seg_1 <= "11111001";
                when X"2" => seven_seg_1 <= "10100100";
                when X"3" => seven_seg_1 <= "10110000";
                when X"4" => seven_seg_1 <= "10011001";
                when X"5" => seven_seg_1 <= "10010010";
                when X"6" => seven_seg_1 <= "10000010";
                when X"7" => seven_seg_1 <= "11111000";
                when X"8" => seven_seg_1 <= "10000000";
                when X"9" => seven_seg_1 <= "10010000";
                when X"A" => seven_seg_1 <= "10001000";
                when X"b" => seven_seg_1 <= "10000011";
                when X"C" => seven_seg_1 <= "11000110";
                when X"d" => seven_seg_1 <= "10100001";
                when X"E" => seven_seg_1 <= "10000110";
                when X"F" => seven_seg_1 <= "10001110";
                when others => seven_seg_1 <= "11111111";
            end case;
        end if;
        --
        if (display_select = '0') then          ③
            seven_seg_2 <= "11111111";              ④
        else
            case(display_cnt_2) is          -- 76543210
                when X"0" => seven_seg_2 <= "11000000";
                when X"1" => seven_seg_2 <= "11111001";
                when X"2" => seven_seg_2 <= "10100100";
                when X"3" => seven_seg_2 <= "10110000";
                when X"4" => seven_seg_2 <= "10011001";
                when X"5" => seven_seg_2 <= "10010010";
                when X"6" => seven_seg_2 <= "10000010";
                when X"7" => seven_seg_2 <= "11111000";
                when X"8" => seven_seg_2 <= "10000000";
                when X"9" => seven_seg_2 <= "10010000";
                when X"A" => seven_seg_2 <= "10001000";
                when X"b" => seven_seg_2 <= "10000011";
                when X"C" => seven_seg_2 <= "11000110";
                when X"d" => seven_seg_2 <= "10100001";
                when X"E" => seven_seg_2 <= "10000110";
                when X"F" => seven_seg_2 <= "10001110";
                when others => seven_seg_2 <= "11111111";
            end case;
        end if;
    end if;
end process;

end architecture Behavioral;
```

1) we choose which display to activate. If "display_select" is one, this means that we want to have the second display active, and so we turn off all the segments of the first one, "seven_seg_1". Noting that the segments of these displays are low-active, we set them all to "1" to turn them off;

2) the "else" here means that "display_select" is zero, and we assign the translated hex values to the first segment, as we saw earlier;

3) if "display_select" is one, we turn off the segments of the second display ("seven_seg_2");

4) otherwise (we're selecting the second segment) we translate the hex values for the second segment.

Digilent "seg_control"

Here's how the 7-segment displays are driven on the Digilent board.

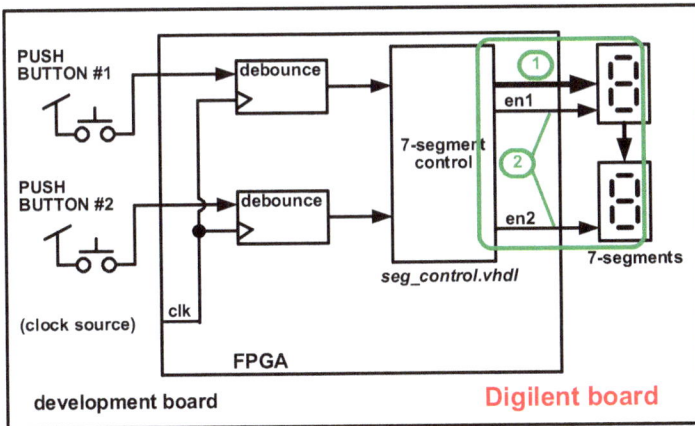

1) one 7-bit display bus is shared by both display units;

2) individual enables select which display is active to display the value on the 7-bit display bus.

Here's the entity declaration:

```
entity seg_control is
    port
    (
        clk             : in    std_logic;
   (1)  pb_debncd_1     : in    std_logic;
        pb_debncd_2     : in    std_logic;
        --
   (2)  seven_seg_1     : out   std_logic_vector(7 downto 0);
   (3)  seven_seg_2     : out   std_logic_vector(7 downto 0); --not used
   (4)  en1             : out   std_logic;
        en2             : out   std_logic
    );
end entity;
```

1) the debounced push-button inputs;
2) the 7-segment bus that sources the external displays;
3) this 7-segment bus is not used on the Digilent board, but since we use the same component instantiation at the top level ("seven_segment.vhdl"), we must include it here;
4) the individual enables for the displays.

The architecture signal declarations:

```
architecture Behavioral of seg_control is
 (1) signal pb_debncd_1_d1 : std_logic;
     signal pb_debncd_2_d1 : std_logic;
 (2) signal display_select : std_logic; -- 0 = #1, 1 = #2
 (3) signal display_cnt_1  : unsigned(3 downto 0);
     signal display_cnt_2  : unsigned(3 downto 0);
 (4) signal display_cnt    : unsigned(3 downto 0);

begin
```

1) used to detect a push-button event;
2) this signal tracks which push-button has last been pushed;
3) these are the display counts; and
4) this is the selected count that is actually displayed.

Here's the architecture body, but don't try to make out the tiny letters – we'll zoom in on the sections:

A) a first process to increment the two counts and toggle the select signal,

B) direct assignments for the display enables, and

C) a second process to implement the case statement display lookup tables as we already saw.

Blaine C. Readler

```
begin

   seg_control_process : process(clk)
   begin
      if rising_edge(clk) then                    (A)
         pb_debncd_1_d1 <= pb_debncd_1;
         pb_debncd_2_d1 <= pb_debncd_2;
         --
         if   (   pb_debncd_1    = '1' --Digilent PBs are high active
               AND pb_debncd_1_d1 = '0'
             ) then
               display_select <= '0';
               display_cnt_1  <= display_cnt_1 + 1;
         elsif (   pb_debncd_2   = '1' --Digilent PBs are high active
               AND pb_debncd_2_d1 = '0'
             ) then
               display_select <= '1';
               display_cnt_2  <= display_cnt_2 + 1;
         end if;
      end if;
   end process;

   en1 <=     display_select; --7-segment enables are low-active.
   en2 <= NOT display_select;
                                                   (B)
   --      0
   --    5   1
   --      6
   --    4   2
   --      3

   seg_decode_process : process(clk)
   begin                                           (C)
      if rising_edge(clk) then
         if (display_select = '0') then
            display_cnt <= display_cnt_1;
         else
            display_cnt <= display_cnt_2;
         end if;
         --   display segments are low-active
         case(display_cnt) is        -- 76543210
            when X"0" => seven_seg_1 <= "11000000";
            when X"1" => seven_seg_1 <= "11111001";
            when X"2" => seven_seg_1 <= "10100100";
            when X"3" => seven_seg_1 <= "10110000";
            when X"4" => seven_seg_1 <= "10011001";
            when X"5" => seven_seg_1 <= "10010010";
            when X"6" => seven_seg_1 <= "10000010";
            when X"7" => seven_seg_1 <= "11111000";
            when X"8" => seven_seg_1 <= "10000000";
            when X"9" => seven_seg_1 <= "10010000";
            when X"A" => seven_seg_1 <= "10001000";
            when X"b" => seven_seg_1 <= "10000011";
            when X"C" => seven_seg_1 <= "11000110";
            when X"d" => seven_seg_1 <= "10100001";
            when X"E" => seven_seg_1 <= "10000110";
            when X"F" => seven_seg_1 <= "10001110";
            when others => seven_seg_1 <= "11111111";
         end case;
      end if;
   end process;
```

76

We'll take each of these in turn.

(A)

```
seg_control_process : process(clk)
begin
   if rising_edge(clk) then                    (1)
       pb_debncd_1_d1 <= pb_debncd_1;
       pb_debncd_2_d1 <= pb_debncd_2;
       --
       if   (    pb_debncd_1     = '1' --Digilent PBs are high-active
            AND pb_debncd_1_d1 = '0'
            ) then                              (2)
            display_select <= '0';
            display_cnt_1  <= display_cnt_1 + 1;
       elsif (    pb_debncd_2     = '1' --Digilent PBs are high-active
            AND pb_debncd_2_d1 = '0'
            ) then
            display_select <= '1';
            display_cnt_2  <= display_cnt_2 + 1;
       end if;
   end if;
end process;
```

This is the first of two processes.

1) we develop a one-clock delayed version of the debounced button signals for each of the two button inputs,

2) which we use to create a one-clock event when a button is pushed. We saw something like this back in the debounce module, but whereas there we detected when the inputs button signal simply changed (went either high or low), here we specifically detect that the input debounced signal goes from low to high (the push-buttons on the Digilent board are high-active) – we find the one clock period where the input has gone high, but its delayed version is still low.

Blaine C. Readler

```
seg_control_process : process(clk)
begin
    if rising_edge(clk) then
        pb_debncd_1_d1 <= pb_debncd_1;
        pb_debncd_2_d1 <= pb_debncd_2;
        --
        if    (    pb_debncd_1    = '1' --Digilent PBs are high-active
              AND pb_debncd_1_d1 = '0'
              ) then
            display_select <= '0';
            display_cnt_1   <= display_cnt_1 + 1;
        elsif (    pb_debncd_2    = '1' --Digilent PBs are high-active
              AND pb_debncd_2_d1 = '0'
              ) then
            display_select <= '1';
            display_cnt_2   <= display_cnt_2 + 1;
        end if;
    end if;
end process;
```

At each of these one-clock button-pushed events,

1) we either clear or set the signal "delay_select" that selects which display to light (low for the first display, and high for the second), and

2) increment the counter associated with the button that was pushed.

Next, we see how we use "delay_select". These segment enables are low-active, so when "delay_select" is low, it enables the first display (via "en1"), and when it's high, it's inverted to enable the second display (via "en2").

Ⓑ

```
en1 <=     display_select; --7-segment enables are low-active.
en2 <= NOT display_select;
```

Here, we see how the hexadecimal counts value from the first process is the source for the ASCII count value displayed.

78

Ⓒ

```
seg_decode_process : process(clk)
begin
   if rising_edge(clk) then
      if (display_select = '0') then
         display_cnt <= display_cnt_1;        ①
      else
         display_cnt <= display_cnt_2;
      end if;                                  ②
      --    display segments are low-active
      case(display_cnt) is          -- 76543210
         when X"0" => seven_seg_1 <= "11000000";
         when X"1" => seven_seg_1 <= "11111001";
         when X"2" => seven_seg_1 <= "10100100";
         when X"3" => seven_seg_1 <= "10110000";
         when X"4" => seven_seg_1 <= "10011001";
         when X"5" => seven_seg_1 <= "10010010";
         when X"6" => seven_seg_1 <= "10000010";
         when X"7" => seven_seg_1 <= "11111000";
         when X"8" => seven_seg_1 <= "10000000";
         when X"9" => seven_seg_1 <= "10010000";
         when X"A" => seven_seg_1 <= "10001000";
         when X"b" => seven_seg_1 <= "10000011";
         when X"C" => seven_seg_1 <= "11000110";
         when X"d" => seven_seg_1 <= "10100001";
         when X"E" => seven_seg_1 <= "10000110";
         when X"F" => seven_seg_1 <= "10001110";
         when others => seven_seg_1 <= "11111111";
      end case;
   end if;
end process;
```

1) we select which count value to apply to the case statement lookup table. This type of selection, choosing among multiple sources for one output, is called a "multiplexer," or just "mux" for short;

2) the mux'd count is now applied to the case lookup table.

Note that in this exercise project we're lighting either one display or the other. How would we use both, or for that matter all four at the same time? The trick is to cycle through all the displays, one after the other, so quickly that the human eye sees them as all active together. This is nicely explained in the Digilent user guide document.

Seven-Segment Action Exercises

Add additional counters and use four of the 7-segment displays. This means driving additional 8-bit buses out of the FPGA for the Terasic board (pin values are shown in the user manual), or for the Digilent board, using all four enable signals. Instead of two pushbuttons, each controlling one display, you could use one button to cycle through the displays, and the other button to increment the count for that display.

Change the case statement to display characters other than the sixteen hex values.

Here's some examples.

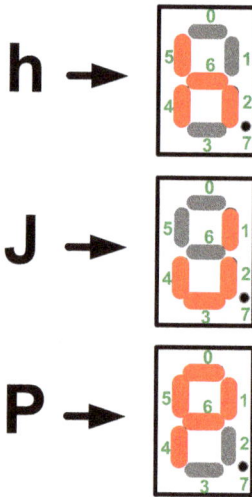

Note that as our projects become more involved, your exercise designs will likely be very different than my examples. This is expected. You might view my examples as additional pieces to the knowledge puzzle.

Chapter Five
Remember This

In this chapter we introduce the use of memories. This project requires a very modest memory that is eight bits wide and eight bytes deep (two of these memories, actually). Memories as small as this are normally implemented with the regular logic resources (specifically, registers). For larger memories, virtually all FPGAs now include built-in memory arrays. These typically consist of memory blocks comprising a fixed number of bits (i.e., cells). We build the memories that we need using one or more of these memory blocks. Each block can be configured to have a variety of widths versus depths.

For example, say we'd like a memory that's 16 bits wide, and 8K deep. The Digilent board's Artix-7 FPGA includes memory blocks that are 36Kb in size. We could use four of their memory blocks, each configured as 8bits wide, 4K deep (with 4Kb left over). The memory would look like this.

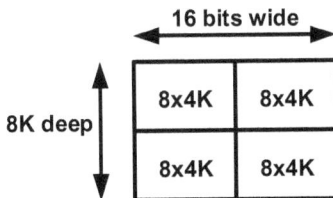

Digilent's Artix-7 FPGA

Alternatively, we could use a different aspect ratio, like this.

16 bits wide
16x2K
16x2K
16x2K
16x2K

8K deep

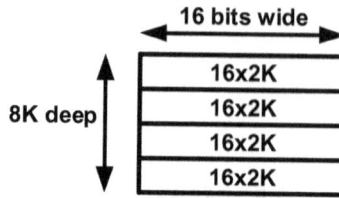

Digilent's Artix-7 FPGA

The Terasic MAX10 FPGA offers memory blocks that are 9Kb, so the memory as configured might look like this.

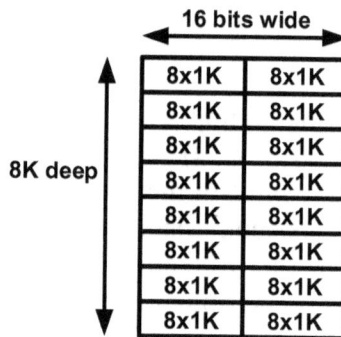

16 bits wide	
8x1K	8x1K
8x1K	8x1K
8x1K	8x1K
8x1K	8x1K
8x1K	8x1K
8x1K	8x1K
8x1K	8x1K
8x1K	8x1K

8K deep

Or this.

16 bits wide

16x512
16x512
16x512
16x512
16x512
16x512
16x512
16x512
16x512
16x512
16x512
16x512
16x512
16x512
16x512

8K deep

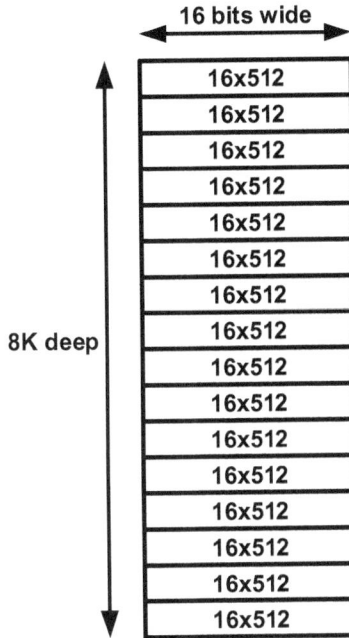

Note that the size of the individual memory blocks is not necessarily indicative of the total available memory of the FPGA, but rather the size coupled with the number of available blocks. Understanding how memories are constructed is useful when gauging whether a projected design will fit in a certain FPGA.

But, on to this chapter's design, where we'll use memories to store display values.

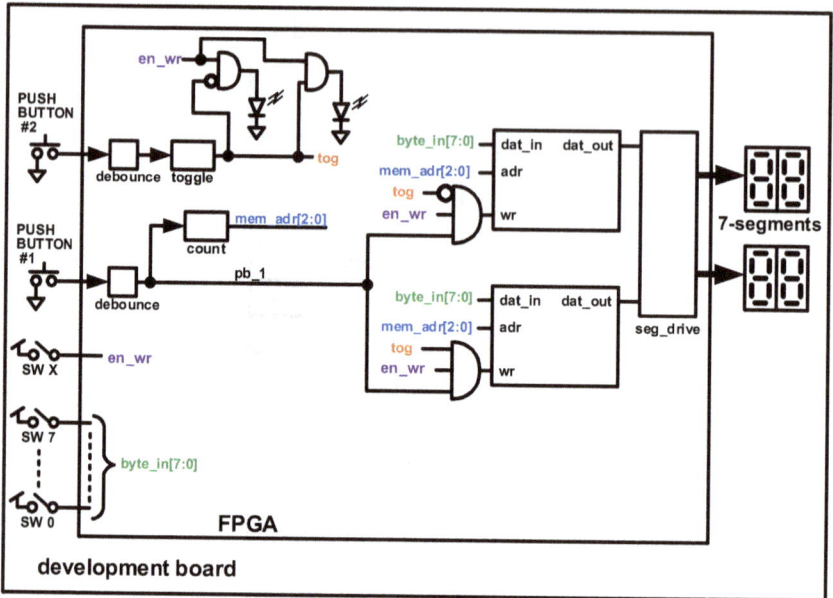

This is the basic flow of information. Note that we're not showing a clock, and going forward we'll assume that this is a given.

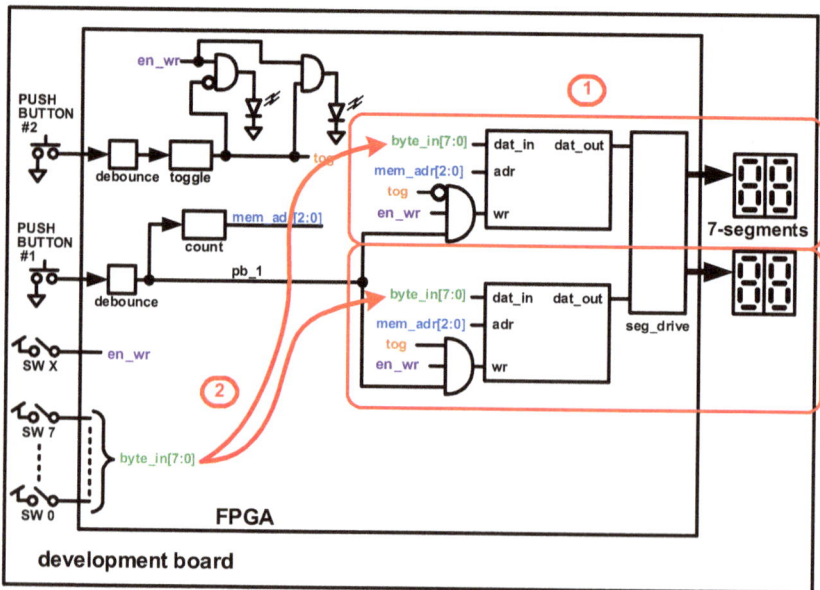

1) one memory for each two-display set, providing a total of four 7-segment display characters;

2) the display characters are entered into the memory using slide switches included on the boards. Note that a four-bit hex value provided by the user is translated into the seven bits that drive a display character. Thus, the eight-bit byte written into each memory is translated (by "seg_drive") into two display characters, one for each display unit, the MS four bits drives one display unit, and the LS four bits drives the other.

The left-most (i.e., MS) slide switch enables writes to the memories (via "en_wr").

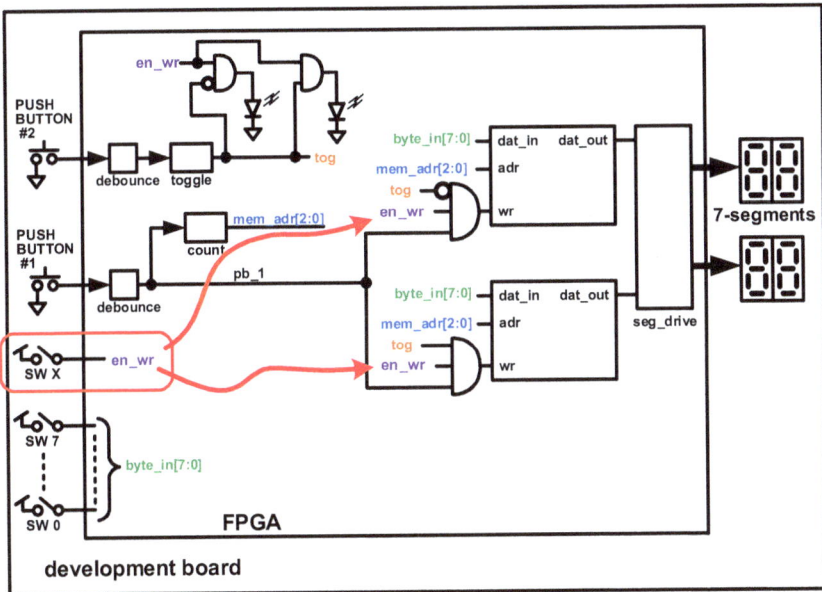

One of the push-buttons selects which memory we write into:

Blaine C. Readler

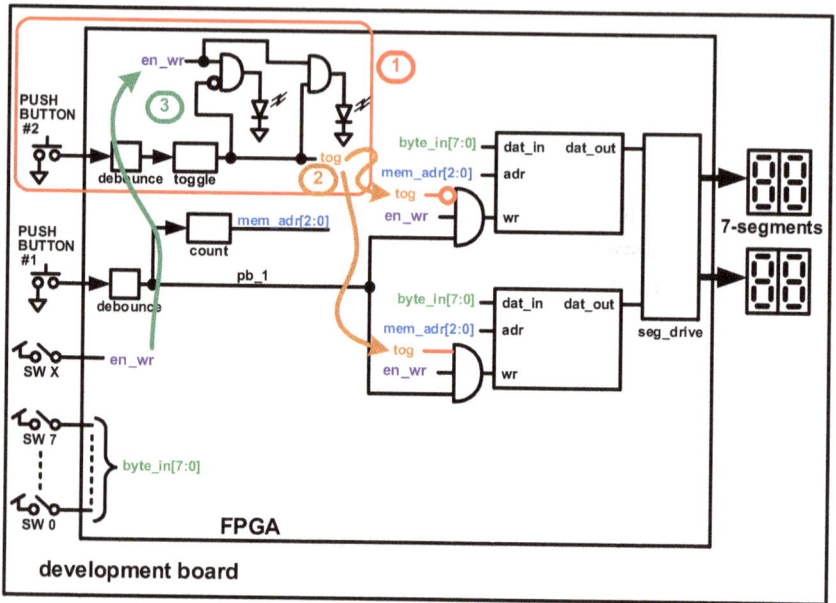

development board

1) each time the second push-button is pressed, the "tog" signal toggles back and forth between high and low,

2) enabling writes to one or the other memory.

3) the "en_wr" slide-switch enables one of two LEDs, which indicate both that the write mode is generally selected, and specifically, which memory is being written.

Pressing the first push-button writes the "byte_in" value to the memory selected by "tog" (via "pb_1"), and increments the memory address ("mem_adr"):

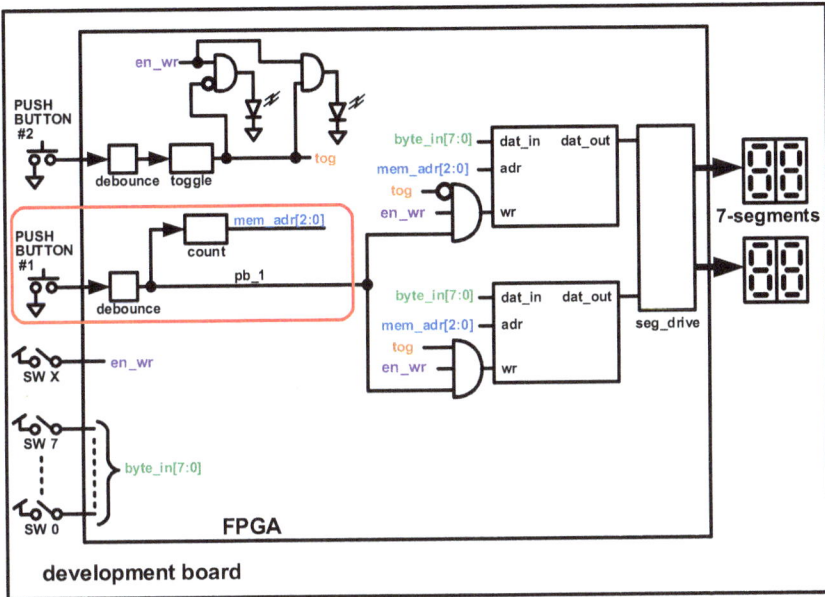

So far we've seen how the writes to the memory operate. Now we'll look at the reads.

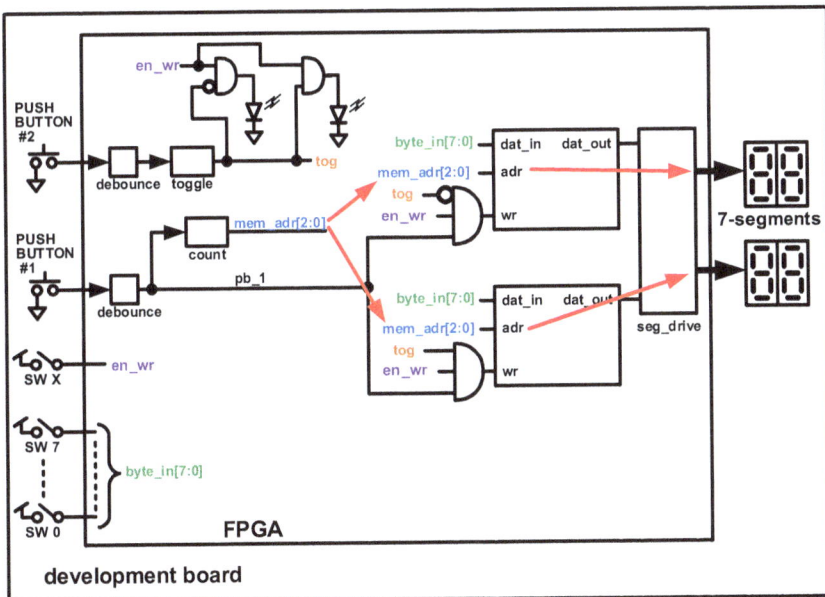

Memory reads are quite straightforward, since the same address used for writing also performs the reads. Thus, when not in write mode (the "en_wr" slide switch is off), each push of the first push-button increments the address, and we see the next stored value that we wrote into memory. Note that the "mem_adr" count rolls over when it reaches seven, so pushing the first push-button cycles through all the memory locations, and then starts over.

The "seg_drive" block translates the four-bit hex values read from memory to the 8-bit characters displayed, but since the operation is different between the two boards (specifically, how the displays are driven), this will be covered separately.

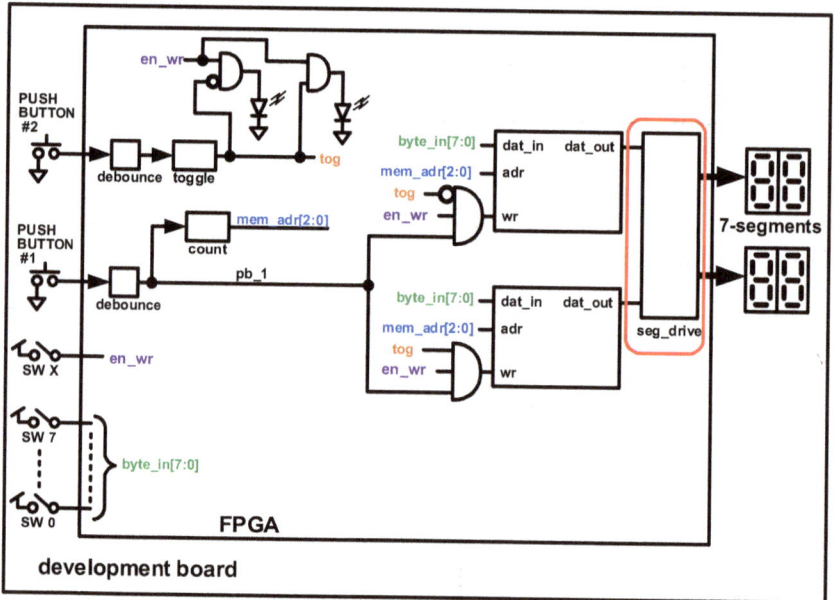

We'll take a moment to look at the various types of memories that we use in FPGAs.

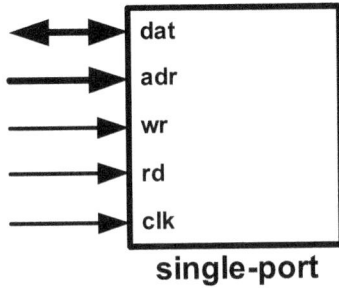

single-port

This is the simplest: one port, one address for both writing and reading. When the "rd" input is low, the data bus points inward, and allows the "wr" input to write data. When the "rd" input is high, the bus turns around, and the memory outputs the addressed data.

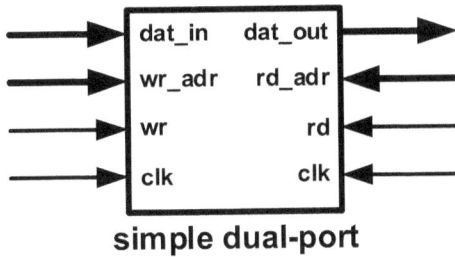

simple dual-port

This is a type of two-port, i.e., dual-port, memory, but the simplest form, where one port is dedicated to writing, and the other to reading. Thus, we have separate address inputs. Importantly, we have two separate clocks, so we can be writing from one section of logic, while simultaneously reading from another – two domains of logic operating off different clocks. The "rd" input is optional; if not present, then the data out always represents the addressed memory location. Note that the two clock inputs can be driven by the same clock, while still allowing simultaneous writing and reading.

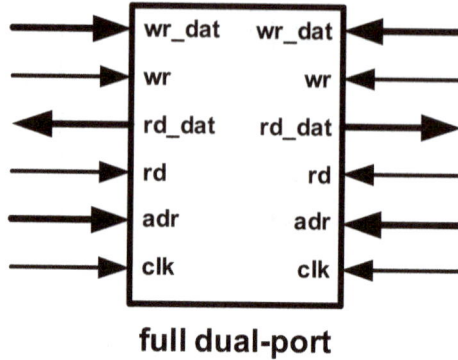

full dual-port

Here we have a full dual-port memory, where we've taken the functionality of the simple dual port, and duplicated it, one for each of the two ports. Both ports can write and read data simultaneously. Note, though, that there is just one internal array of memory cells. The two ports are accessing the same memory, able to pass data back and forth.

You can see that our project memory is essentially the simple dual-port with the clocks and write and read addresses tied together:

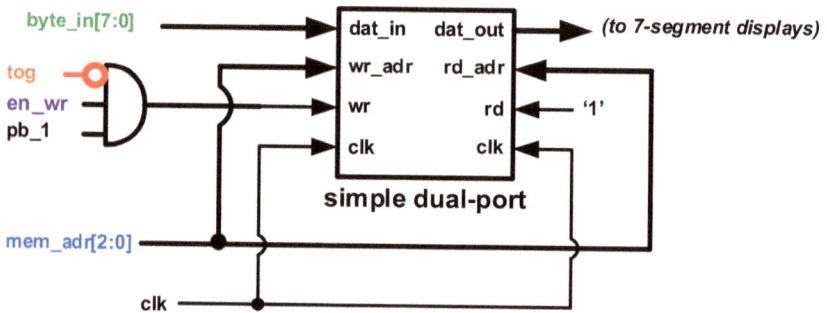

We'll now look at the VHDL code that implements our memory-based design. As our projects are progressing and taking on more complexity, we review the code in functional sections.

memories.vhdl

entity declaration

component declarations

signal declarations

debounce instantiations

main process

continuous assignments

memory instantiations

seg_drive instantiation

body

architecture

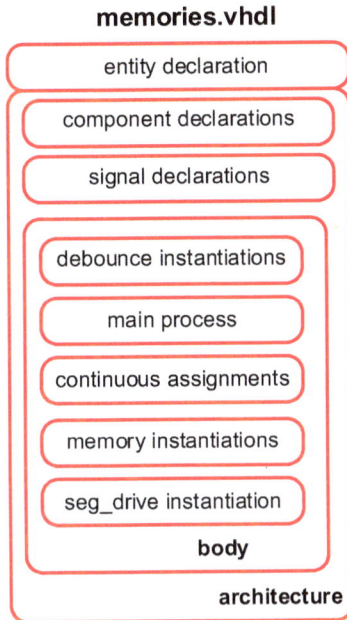

This is the breakdown of the top project file, "memories.vhdl".

Here's the entity declaration.

memories.vhdl

entity declaration

component declarations

signal declarations

debounce instantiations

main process

continuous assignments

memory instantiations

seg_drive instantiation

body

architecture

```
entity memories is
  port
  (
    clk          : in    std_logic;        (1)
    pb_1         : in    std_logic;
    pb_2         : in    std_logic;        (2)
(3) en_wr        : in    std_logic;
    sw           : in    std_logic_vector(7 downto 0) ; (4)
    --
    seven_seg_1  : out   std_logic_vector(7 downto 0) ; (5)
    seven_seg_2  : out   std_logic_vector(7 downto 0) ;
    seven_seg_3  : out   std_logic_vector(7 downto 0) ; (6)
    seven_seg_4  : out   std_logic_vector(7 downto 0) ;
    en1          : out   std_logic;
    en2          : out   std_logic;        (7)
    en3          : out   std_logic;
    en4          : out   std_logic;
    led_1        : out   std_logic;        (8)
    led_2        : out   std_logic;
    -- unused LEDs                         (9)
    led          : out   std_logic_vector(9 downto 2)
  ) ;
end entity;
```

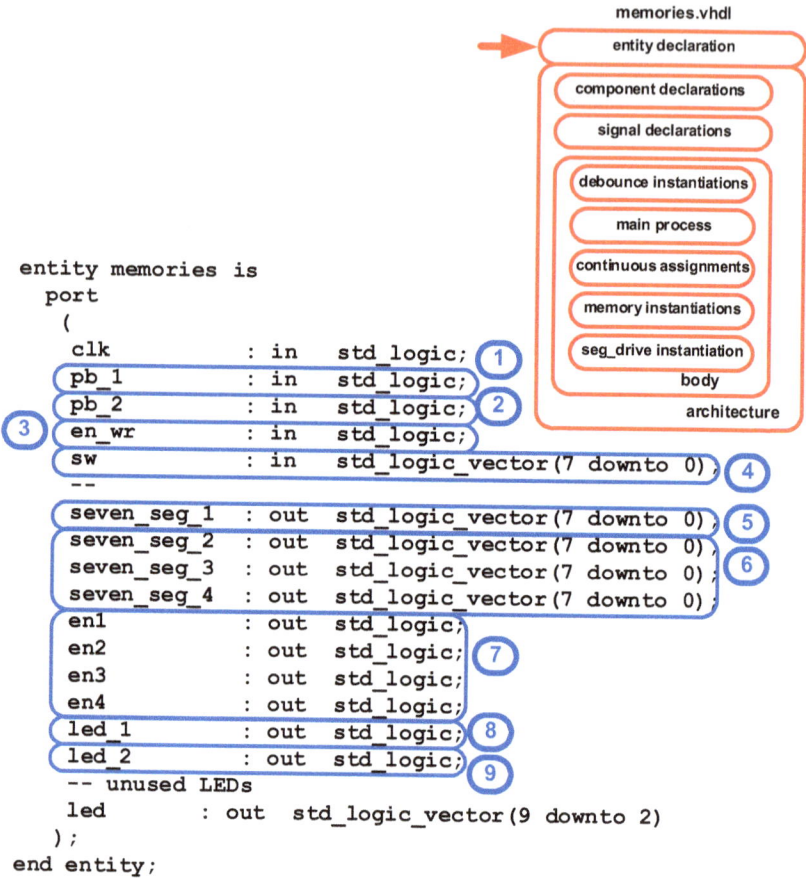

1) "pb_1" writes the slide-switch byte to memory and increments the memory address;

2) "pb_2" toggles which memory is written (when the slide-switch in set to write mode);

3) "en_wr" is the slide-switch which selects either write mode (when set), or read mode;

4) "sw" are the slide-switches which establish the byte to be written;

5) first 7-segment output – the first display component for the Terasic board, and the only one for the Digilent board;

6) the rest of the display components for the Terasic board, and not used for the Digilent;

7) the continuously cycling 7-segment enables for the Digilent board, and not used for the Terasic board;

8) when lit, "led_1" indicates that writes are made to the memory associated with the two 7-segment display components on the left;

9) when lit, "led_2" indicates that writes are made to the memory associated with the two 7-segment display components on the right;

Next are the component declarations:

```
architecture Behavioral of memories is

  -- component declarations
```

memories.vhdl

- entity declaration
- component declarations
- signal declarations
- debounce instantiations
- main process
- continuous assignments
- memory instantiations
- seg_drive instantiation

body

architecture

```
component debounce is                          1
port
  (
  clk              : in    std_logic;
  pb_in            : in    std_logic;
  --
  pb_debounced : out   std_logic
  );
  end component;
```

```
component memory_8x8
port                                           2
  (
  clk          : in  std_logic;
  dat_in       : in  std_logic_vector (7 downto 0);
  adr          : in  std_logic_vector (2 downto 0);
  wr           : in  std_logic;
  dat_out      : out std_logic_vector (7 downto 0)
  );
end component;
```

```
component seg_drive
port                                           3
  (
  clk          : in    std_logic;
  mem_dat_1    : in    std_logic_vector (7 downto 0);
  mem_dat_2    : in    std_logic_vector (7 downto 0);
  --
  seven_seg_1  : out   std_logic_vector (7 downto 0);
  seven_seg_2  : out   std_logic_vector (7 downto 0);
  seven_seg_3  : out   std_logic_vector (7 downto 0);
  seven_seg_4  : out   std_logic_vector (7 downto 0);
  en1          : out   std_logic;
  en2          : out   std_logic;
  en3          : out   std_logic;
  en4          : out   std_logic
  );
end component;
```

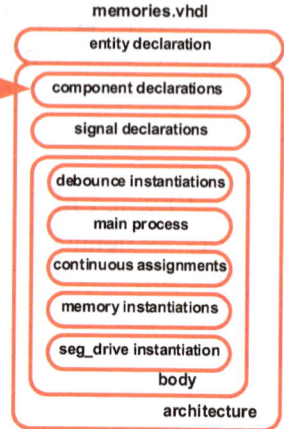

1) the debounce module that we've seen before;

2) the memory – two total, instantiated later;

3) the "segment drive" module – like "seg_control" from the 7-segment project, "seg_drive" is different between the two

development boards, and so the vendor projects (Quartus and Vivado) each point to their own version.

Signal declarations, used later in the architecture body:

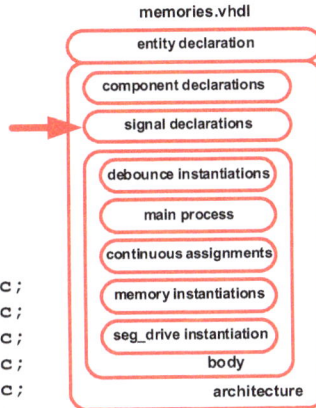

```
-- signal declarations
signal pb_debncd_1      : std_logic;
signal pb_debncd_1_d1   : std_logic;
signal pb_debncd_2      : std_logic;
signal pb_debncd_2_d1   : std_logic;
signal tog              : std_logic;
signal pb1_pulse        : std_logic;
signal mem_adr          : unsigned(2 downto 0);
signal byte_in          : std_logic_vector(7 downto 0);
signal mem_adr_std      : std_logic_vector(2 downto 0);
signal mem_wr_1         : std_logic;
signal mem_wr_2         : std_logic;
signal mem_dat_out_1    : std_logic_vector(7 downto 0);
signal mem_dat_out_2    : std_logic_vector(7 downto 0);
```

We begin the architecture body with the two debounce module instantiations. Again, we've seen these before:

Blaine C. Readler

memories.vhdl

entity declaration

component declarations

signal declarations

← debounce instantiations

main process

continuous assignments

memory instantiations

seg_drive instantiation

body

architecture

```
begin

    debounce_1 : debounce
    port map
      (
      clk            => clk,        --in    std_logic;
      pb_in          => pb_1,       --in    std_logic;
      --
      pb_debounced => pb_debncd_1  --out   std_logic
      );

    debounce_2 : debounce
    port map
      (
      clk            => clk,        --in    std_logic;
      pb_in          => pb_2,       --in    std_logic;
      --
      pb_debounced => pb_debncd_2  --out   std_logic
      );
```

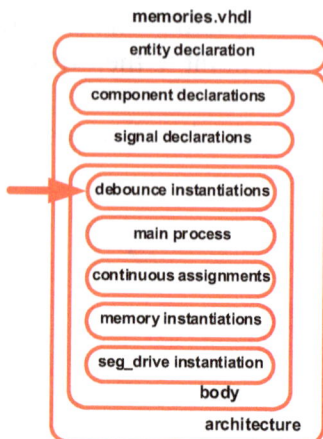

This is the clocked process statement, including multiple functions:

```
main_proc : process(clk)
begin
    if rising_edge(clk) then
        pb_debncd_2_d1 <= pb_debncd_2;
        if (pb_debncd_2 = '0' and pb_debncd_2_d1 = '1') then
            tog <= not tog;                                    1
        end if;
        --
        pb_debncd_1_d1 <= pb_debncd_1;
        if (pb_debncd_1 = '0' and pb_debncd_1_d1 = '1') then
            mem_adr <= mem_adr + 1;              4            3
            pb1_pulse <= '1';
        else
            pb1_pulse <= '0';     5
        end if;
    end if;
end process;
```

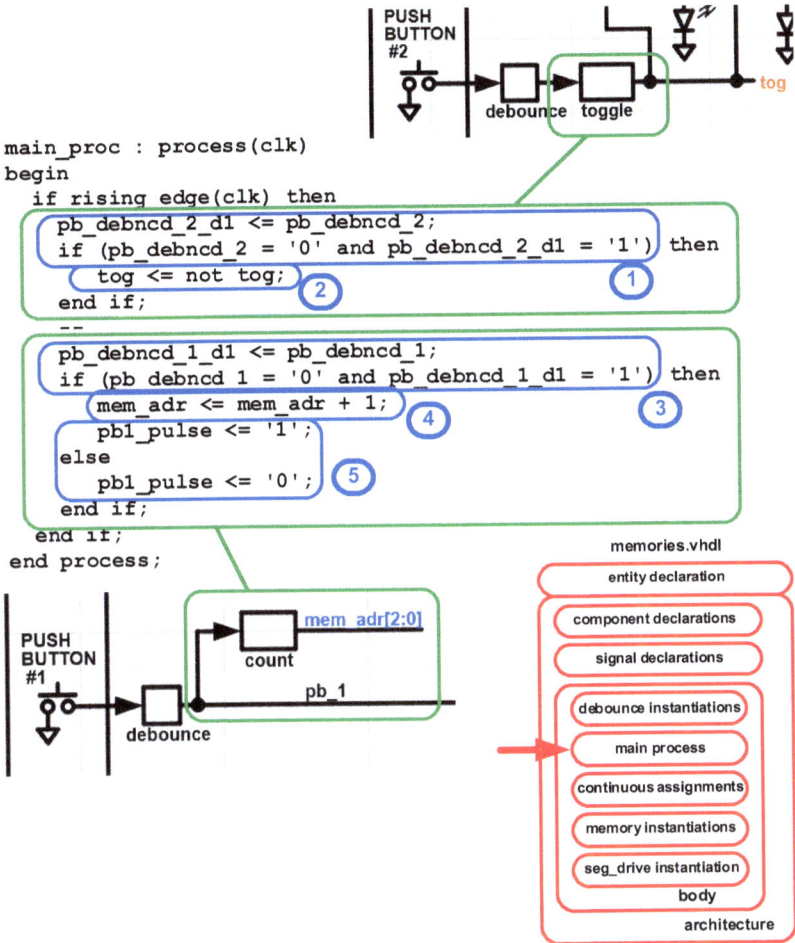

1) as we've seen before, this detects the falling edge of the debounced push-button #2;

2) which toggles the "tog" signal;

3) once more we detect the falling edge, this time of the debounced push-button #1;

4) which increments the memory address, and

5) pulses "pb1_pulse" for one clock.

Next are direct assignments of signals. Since the assignments do not occur at one instant, as in a clocked process, we call these "continuous" assignments.

97

```
led_1 <= en_wr and not tog;
led_2 <= en_wr and     tog;
```

```
byte_in     <= sw(7 downto 0);
mem_wr_1    <= (not tog) and en_wr and pb1_pulse;
mem_wr_2    <=      tog  and en_wr and pb1_pulse;
mem_adr_std <= std_logic_vector(mem_adr);
```

memories.vhdl

entity declaration
component declarations
signal declarations
debounce instantiations
main process
continuous assignments
memory instantiations
seg_drive instantiation
body
architecture

1) this is self-explanatory;

2) a simple renaming for convenience;

3) the logic gating to produce the memory writes;

4) here, we're taking care of a VHDL-specific quirk. In order to create a counter that increments via arithmetic (adding one each count), we had to declare "mem_adr" as "unsigned." However, signals that enter/exit files generally must be "std_logic" or "std_logic_vector" types, and this line simply converts the unsigned "mem_adr" to the std_logic_vector "mem_adr_std" so that we can connect it to the memory modules.

The two memory module instantiations:

98

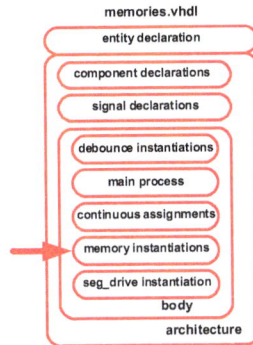

```
memory_8x8_1 : memory_8x8
port map
  (
  clk     => clk,          --in  std_logic;
  dat_in  => byte_in,      --in  std_logic_vector (7 downto 0);
  adr     => mem_adr_std,  --in  std_logic_vector (2 downto 0);
  wr      => mem_wr_1,     --in  std_logic;
  dat_out => mem_dat_out_1 --out std_logic_vector (7 downto 0)
  );
```

```
memory_8x8_2 : memory_8x8
port map
  (
  clk     => clk,          --in  std_logic;
  dat_in  => byte_in,      --in  std_logic_vector (7 downto 0);
  adr     => mem_adr_std,  --in  std_logic_vector (2 downto 0);
  wr      => mem_wr_2,     --in  std_logic;
  dat_out => mem_dat_out_2 --out std_logic_vector (7 downto 0)
  );
```

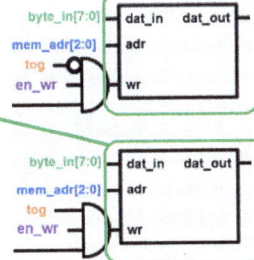

And finally, the "seg_drive" instantiation.

99

Blaine C. Readler

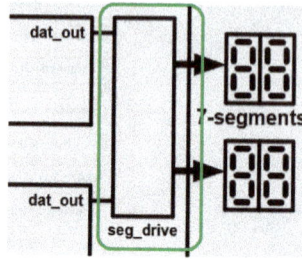

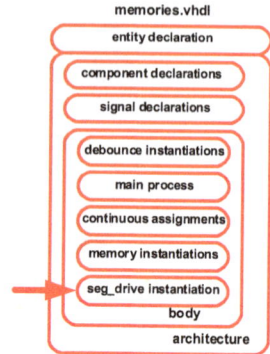

```
seg_drive_i : seg_drive
port map
  (
  clk          => clk,              --in    std_logic;
  mem_dat_1    => mem_dat_out_1,    --in    std_logic_vector(7 downto 0);
  mem_dat_2    => mem_dat_out_2,    --in    std_logic_vector(7 downto 0);
  --
  seven_seg_1  => seven_seg_1,      --out   std_logic_vector(7 downto 0);
  seven_seg_2  => seven_seg_2,      --out   std_logic_vector(7 downto 0);
  seven_seg_3  => seven_seg_3,      --out   std_logic_vector(7 downto 0);
  seven_seg_4  => seven_seg_4,      --out   std_logic_vector(7 downto 0);
  en1          => en1,              --out   std_logic;
  en2          => en2,              --out   std_logic;
  en3          => en3,              --out   std_logic;
  en4          => en4               --out   std_logic
  );

end architecture Behavioral;
```

Next we'll dive into the memory module, "memory_8x8.vhdl", where we are introduced to VHDL arrays. We can think of a two-dimensional array as a table of register signals which we can point to with an index value. If we think of each register signal as a location in a memory, and the index value as the memory address, then we can see that an array can directly implement a memory.

Here's the code for the memory module:

100

```
entity memory_8x8 is
port
 (
  clk          : in  std_logic;
  dat_in       : in  std_logic_vector (7 downto 0);
  adr          : in  std_logic_vector (2 downto 0);
  wr           : in  std_logic;
  dat_out      : out std_logic_vector (7 downto 0)
 );
 end entity;

architecture Behavioral of memory_8x8 is
                                                        1
    type array_8x8 is array (0 to 7) of std_logic_vector (7 downto 0);
    signal mem : array_8x8;           2
    --
    signal adr_int : integer;    3

begin

    adr_int <= to_integer (unsigned (adr));  4

    main_proc : process (clk)
    begin
      if rising_edge (clk) then          5
        if (wr = '1') then         6
          mem (adr_int) <= dat_in;
        end if;
      end if;
      --
      dat_out <= mem (adr_int);   7
    end process;

end architecture Behavioral;
```

1) we have already seen that VHDL defines signals as different types – so far we've seen both std_logic_vectors and unsigned numbers. We'll soon add integers to the list. However, VHDL also allows us to make up our own signal types, and that's exactly what we'll do to create an array for our memory. We create a new signal type with the keyword "type" followed by the type of name we choose – here we've chosen "array_8x8", followed by the keyword "is". In this case, the new signal type is an array, and we define its size with "0 to 7". We could have created, for example, a hundred-deep array with "1 to 100". After the keyword "of" we must define what the signals of the array consist of, in our case, std_logic_vector. We could have, for example, defined the array to be of unsigned values, or even integers;

2) once we've defined our new signal type, we can declare a signal called "mem" as this new type. Understand, "mem" is now an array of std_logic_vector values;

3) the array "mem" requires its index values to be integers, so we declare "adr_int" as such – this will become our memory address;

4) the address input to our module, "adr", is of the std_logic_vector type, and we need to convert that to the integer type of "adr_int". This is done in a somewhat roundabout way here. Inside the inner parenthesis, the keyword "unsigned" first converts the std_logic_vector "adr" to an unsigned type, and then the outer parenthesis converts that finally to an integer with "to_integer";

5) this clocked process is the memory, where,

6) when the write enable "wr" is active, the input data "dat_in" is written into the array location – i.e., the memory location – as defined by the converted input address "adr_int", and

7) reads are performed each clock cycle as the converted input address "adr_int" accesses a byte from the array, AKA memory.

The last module is "seg_drive", where, like "seg_control" of the previous project, the differences between how the two development boards drive their 7-segment displays requires that each vendor project (Quartus and Vivado) point to their own version.

First, though, we'll look at a sub-module common to both versions of "seg_drive" – the translation from a four-bit hex value to the 8-bit 7-segment drive signal.

```
entity seg_encode is
  port
    (
      hex_in      : in    std_logic_vector(3 downto 0);
      seven_seg   : out   std_logic_vector(7 downto 0)
    );
end entity;

architecture Behavioral of seg_encode is

begin                                    ( 1 )

    seg_decode_process : process(hex_in) ( 2 )
    begin
      case(hex_in) is              -- 76543210
        when X"0" => seven_seg <= "11000000";
        when X"1" => seven_seg <= "11111001";
        when X"2" => seven_seg <= "10100100";
        when X"3" => seven_seg <= "10110000";
        when X"4" => seven_seg <= "10011001";
        when X"5" => seven_seg <= "10010010";
        when X"6" => seven_seg <= "10000010";
        when X"7" => seven_seg <= "11111000";
        when X"8" => seven_seg <= "10000000";
        when X"9" => seven_seg <= "10010000";
        when X"A" => seven_seg <= "10001000";
        when X"b" => seven_seg <= "10000011";
        when X"C" => seven_seg <= "11000110";
        when X"d" => seven_seg <= "10100001";
        when X"E" => seven_seg <= "10000110";
        when X"F" => seven_seg <= "10001110";
        when others => seven_seg <= "11111111";
      end case;
    end process;

end architecture Behavioral;
```

1) this looks like the same case statement we used in the 7-segment project, where the case statement acts as a translation lookup table. There is a fundamental difference, though. Whereas the case statement of the 7-segment project resided inside a clocked process, this one does not. Instead, without a rising-edge clock, the statement works as a continuous assignment;

2) the contents of the parenthesis following the keyword "process" is called the "sensitivity list," and defines when the process should be active – in other words, what are the inputs to the process that affect its outputs (the assigned signals) in some way. You can see that when the only sensitivity entry is the clock, and the internal "IF" statement is qualified with just the

rising edge of that clock, we have a clocked register, where the definition of a register is that the outputs change precisely at the clock's edge. In this case, however, the process output – the "seven_seg" signal – can change with any changes of the (only) input, "hex_in". In our case, we have just one input signal ("hex_in"), but if there were others, we would need to include them in the sensitivity list (although a lot of compilers now might do that for you with a warning shake of the metaphorical digital finger).

We'll see how this sub-module is used in the two versions of "seg_drive", but first we need to take a look at a new type of VHDL statement. This is used in the Digilent "seg_drive", and we look at it here as well for the Digilent users who otherwise wouldn't be enlightened. This is the VHDL "When/Else" statement.

```
WHEN/ELSE
out_1 <= '1' when sel = "00" else '0';
out_2 <= '1' when sel = "01" else '0';
out_3 <= '1' when sel = "10" else '0';
out_4 <= '1' when sel = "11" else '0';
```

The operation is straightforward: the first parameter ('1' here) is assigned to the statement signal ("out_1") when the first condition is met (sel = "00"), otherwise the assignment comes from the last parameter ('0').

Instead of simple '0' and '1' logic levels, the parameters could be other signals. Like so.

```
out_1 <= sig_a when sel = "00" else sig_b;
```

We can also string these together to select from among different signals to assign.

```
out_1 <= sig_a when sel = "00" else
         sig_b when sel = "01" else
         sig_c when sel = "10" else
         sig_d;
```

In this case, you can see that we've implemented a multiplexer (a mux).

Terasic "seg_drive"

Continuing our new method of looking at sections of VHDL code, here's the Terasic board "seg_drive" entity declaration.

```
entity seg_drive is
  port
    (
    clk          : in   std_logic;
    mem_dat_1    : in   std_logic_vector(7 downto 0);
    mem_dat_2    : in   std_logic_vector(7 downto 0);
    --
    seven_seg_1  : out  std_logic_vector(7 downto 0);
    seven_seg_2  : out  std_logic_vector(7 downto 0);
    seven_seg_3  : out  std_logic_vector(7 downto 0);
    seven_seg_4  : out  std_logic_vector(7 downto 0);
    en1          : out  std_logic;
    en2          : out  std_logic;
    en3          : out  std_logic;
    en4          : out  std_logic
    );
  end entity;
```

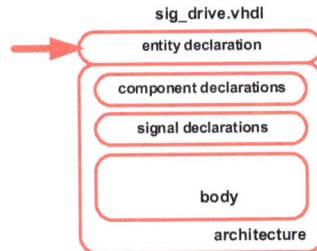

sig_drive.vhdl

entity declaration

component declarations

signal declarations

body

architecture

We note that for the Terasic board, we have separate 7-segment outputs for each of the four (out of six total) displays that we're using for this project.

We have just one sub-module, the "seg_encode" that we just saw, and that are shared across both the Digilent and Terasic boards.

105

```
architecture Behavioral of seg_drive is

  component seg_encode
  port
   (
    hex_in     : in    std_logic_vector(3 downto 0);
    seven_seg  : out   std_logic_vector(7 downto 0)
   );
  end component;

begin
```

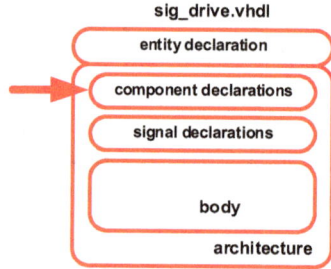

sig_drive.vhdl

entity declaration

→ component declarations

signal declarations

body

architecture

Within the architecture body, we find the four "seg_encode" module instantiations:

sig_drive.vhdl

entity declaration

component declarations

signal declarations

body

architecture

```
begin

  en1 <= '0'; --fixed active for Terasic
  en2 <= '0'; -- "
  en3 <= '0'; -- "
  en4 <= '0'; -- "

  seg_encode_1 : seg_encode
  port map
    (
     hex_in     => mem_dat_1(7 downto 4),
     seven_seg  => seven_seg_1
    );

  seg_encode_2 : seg_encode
  port map
    (
     hex_in     => mem_dat_1(3 downto 0),
     seven_seg  => seven_seg_2
    );

  seg_encode_3 : seg_encode
  port map
    (
     hex_in     => mem_dat_2(7 downto 4),
     seven_seg  => seven_seg_3
    );

  seg_encode_4 : seg_encode
  port map
    (
     hex_in     => mem_dat_2(3 downto 0),
     seven_seg  => seven_seg_4
    );

end architecture Behavioral;
```

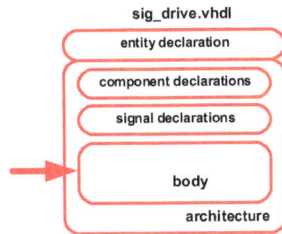

And, we see that each "seg_drive" module controls one 7-segment display, where,

107

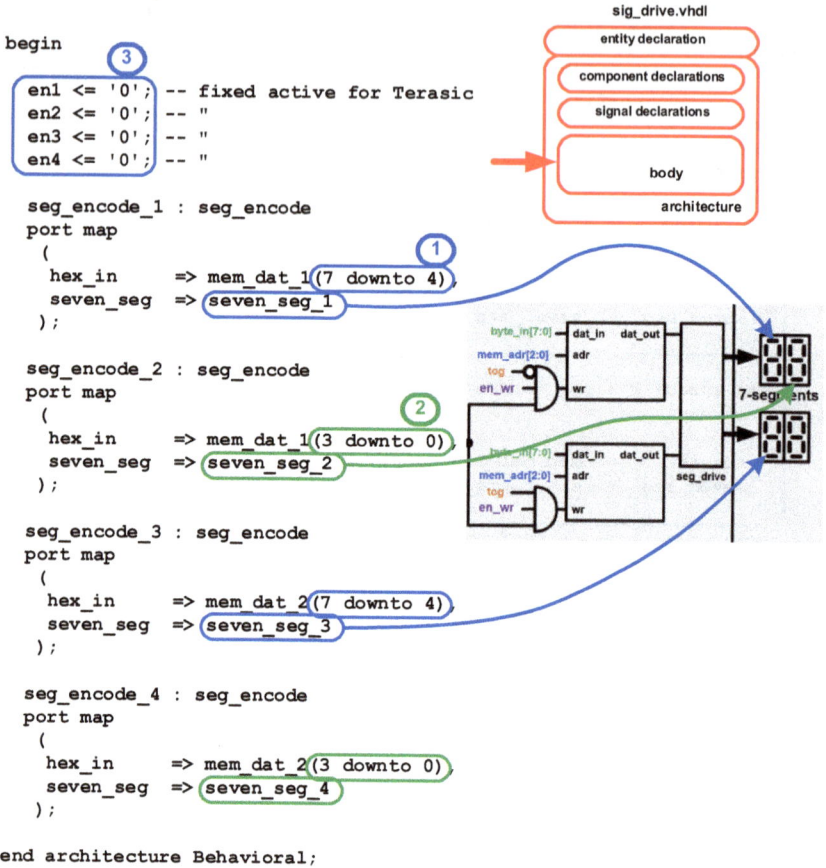

```
begin
   ③
   en1 <= '0';  -- fixed active for Terasic
   en2 <= '0';  -- "
   en3 <= '0';  -- "
   en4 <= '0';  -- "

   seg_encode_1 : seg_encode
   port map
     (
      hex_in      => mem_dat_1(7 downto 4),
      seven_seg   => seven_seg_1
     );

   seg_encode_2 : seg_encode
   port map
     (
      hex_in      => mem_dat_1(3 downto 0),
      seven_seg   => seven_seg_2
     );

   seg_encode_3 : seg_encode
   port map
     (
      hex_in      => mem_dat_2(7 downto 4),
      seven_seg   => seven_seg_3
     );

   seg_encode_4 : seg_encode
   port map
     (
      hex_in      => mem_dat_2(3 downto 0),
      seven_seg   => seven_seg_4
     );

end architecture Behavioral;
```

1) the MS nibble from the memory controls the left-most display character,

2) while the LS nibble from the memory controls the right-most character;

3) for the Terasic board, the four left-most 7-segment displays are always enabled, and thus fixed at zero since they are low-active.

Here's the operation of the Terasic board for this project.

1) the left-most switch, when off (towards the edge of the board) places the board in read mode, while when on (towards the displays), places the board in write mode;

2) these four slide switches set the display hex character for either display characters "A" or "C" (as later retrieved from memory), depending which memory is selected;

3) these four slide switches set the display hex character for either display characters "B" or "D" (as later retrieved from memory), depending which memory is selected;

4) when in a write mode, this LED is lit when the entries are being written to the first memory (associated with displays "A" and "C");

5) when in a write mode, this LED is lit when the entries are being written to the second memory (associated with displays "B" and "D"). Note that when in read mode, neither LED is lit;

6) this push-button toggles between the two memories when in write mode, i.e., toggles between the two LEDs;

7) when in write mode, this push-button writes the selected slide-switch nibble value into its associated memory, and increments the memory address. When in read mode, this push-button simply increments the memory address for the next read value.

109

Digilent "seg_drive"

Staying with our new method of looking at VHDL code in sections, here's the Digilent board "seg_drive" entity declaration.

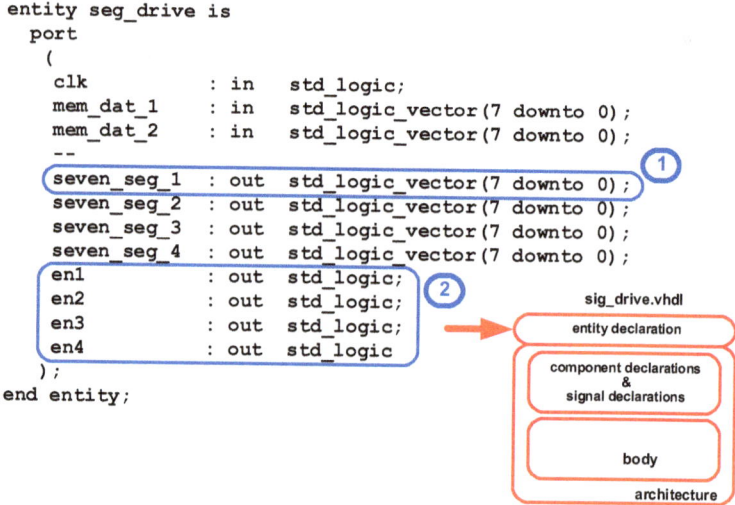

```
entity seg_drive is
  port
    (
     clk            : in    std_logic;
     mem_dat_1      : in    std_logic_vector(7 downto 0);
     mem_dat_2      : in    std_logic_vector(7 downto 0);
     --
     seven_seg_1    : out   std_logic_vector(7 downto 0);     ①
     seven_seg_2    : out   std_logic_vector(7 downto 0);
     seven_seg_3    : out   std_logic_vector(7 downto 0);
     seven_seg_4    : out   std_logic_vector(7 downto 0);
     en1            : out   std_logic;
     en2            : out   std_logic;                        ②
     en3            : out   std_logic;
     en4            : out   std_logic
    );
  end entity;
```

sig_drive.vhdl

entity declaration

component declarations
&
signal declarations

body

architecture

1) for the Digilent board we use just one 7-segment output for all four displays;

2) and all four display enables then select among the four.

Component and signal declarations come next.

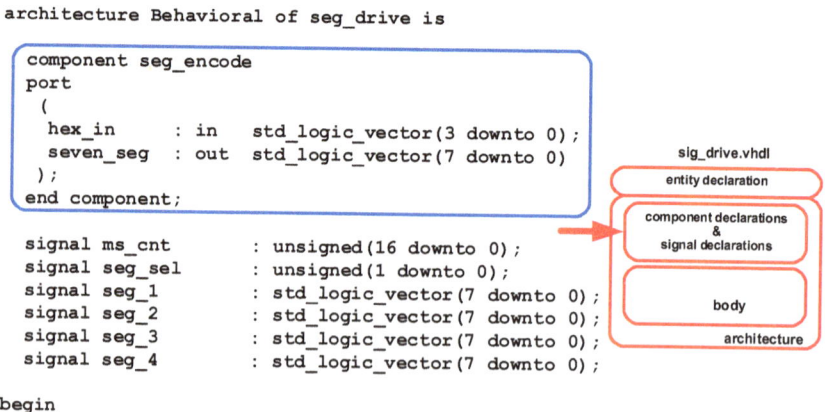

```
architecture Behavioral of seg_drive is

  component seg_encode
  port
    (
     hex_in     : in    std_logic_vector(3 downto 0);
     seven_seg  : out   std_logic_vector(7 downto 0)
    );
  end component;

  signal ms_cnt      : unsigned(16 downto 0);
  signal seg_sel     : unsigned(1 downto 0);
  signal seg_1       : std_logic_vector(7 downto 0);
  signal seg_2       : std_logic_vector(7 downto 0);
  signal seg_3       : std_logic_vector(7 downto 0);
  signal seg_4       : std_logic_vector(7 downto 0);

begin
```

sig_drive.vhdl

entity declaration

component declarations
&
signal declarations

body

architecture

110

We have only one sub-module, the "seg_encode" that we just saw.

On to the body.

Each encoding module translates one nibble from one of the memories to drive one of the 7-segment displays;

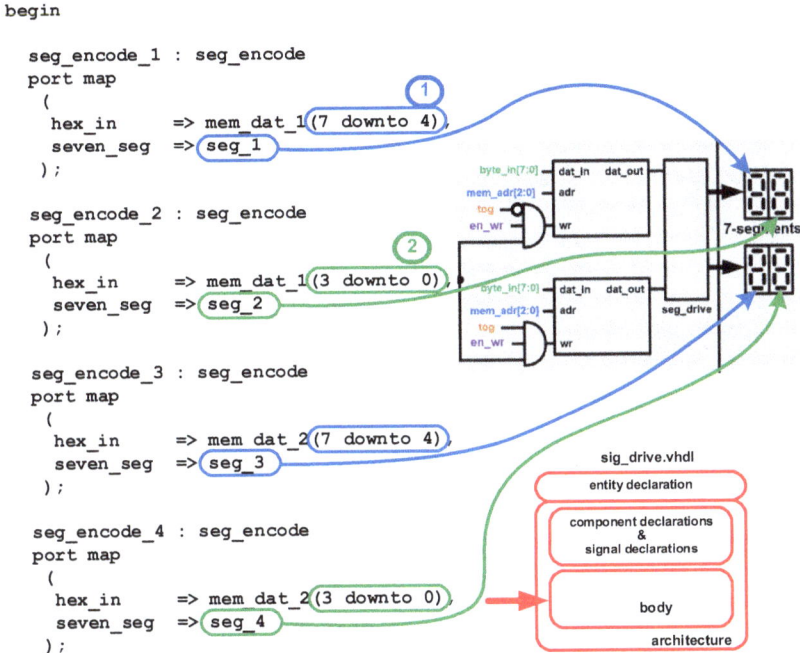

1) the MS nibble from the memory controls the left-most display character,

2) while the LS nibble from the memory controls the right-most character.

As we saw in the 7-segment project, the four 7-segment displays on the Digilent board share one 8-bit display drive bus (here, seven_seg_1), and each display is selected by its own enable (en1, en2, en3, en4). The rest of the architecture body cycles around the four displays, enabling each in turn.

```
-- Cycle through the four segments.
main_proc : process(clk)
begin
   if rising edge(clk) then  (1)  unsigned(16 downto 0)
      ms_cnt <= ms_cnt + 1;  -- ~~1.3ms at 100MHz      sig_drive.vhdl

      if (ms_cnt = '0' & X"0000") then  (2)
         seg_sel <= seg_sel + 1;
      end if;            unsigned(1 downto 0)

   end if;
end process;                (3)

-- 7-segment enables are low-active
en1 <= '0' when seg_sel = "00" else '1';
en2 <= '0' when seg_sel = "01" else '1';
en3 <= '0' when seg_sel = "10" else '1';
en4 <= '0' when seg_sel = "11" else '1';   (4)

seven_seg_1 <= seg_1 when seg_sel = "00" else
               seg_2 when seg_sel = "01" else
               seg_3 when seg_sel = "10" else
               seg_4;

seven_seg_2 <= X"00";  (5)
seven_seg_3 <= X"00";
seven_seg_4 <= X"00";

end architecture Behavioral;
```

Diagram labels: sig_drive.vhdl; entity declaration; component declarations & signal declarations; body; architecture

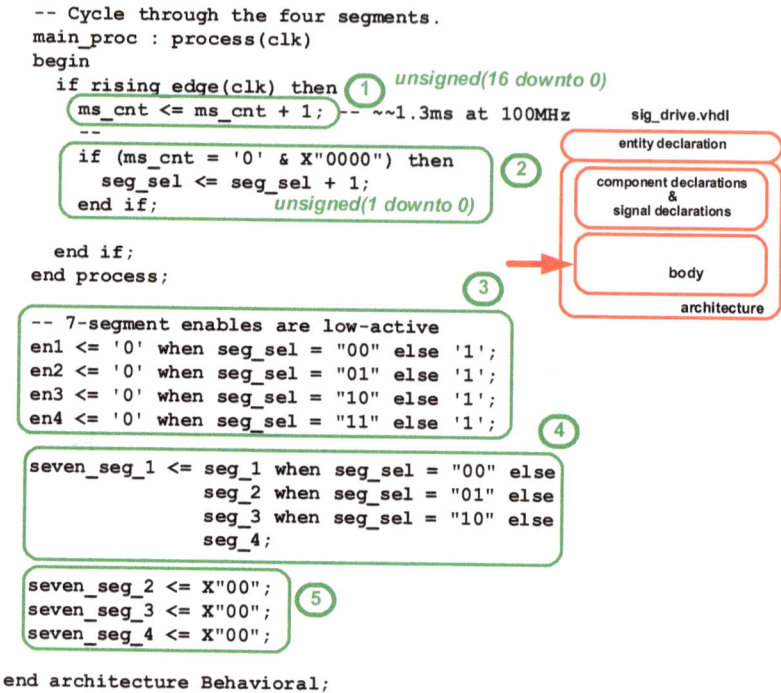

1) the counter, "ms_cnt", continually increments, cycling from 0x00 through 0x1FFF (0 to 131,071), and then starts over. At the 100MHz clock rate, the complete cycle time is about 1.3 milliseconds;

2) each time the 1.3ms cycle starts over (passes through zero) the 2-bit "seg_sel" signal increments, counting from "00" to "11", also starting over when it reaches "11";

3) as seg_sel cycles through its four counts, it activates the four displays in turn, noting that the display enables are low-active;

4) at the same time, the translated hex values from the memories (seg_1, etc.) are mux'd (selected) to the shared bus, "seven_seg_1". Thus, the contents of the two memories are presented to the 7-segment displays in turn, and each display is concurrently enabled. As the Digilent user manual explains, the human eye perceives the four displays as though they are continually lit, each with their own value;

5) since the Digilent board uses a single shared display bus, the others are tied off.

Here's the operation of the Digilent board for this project.

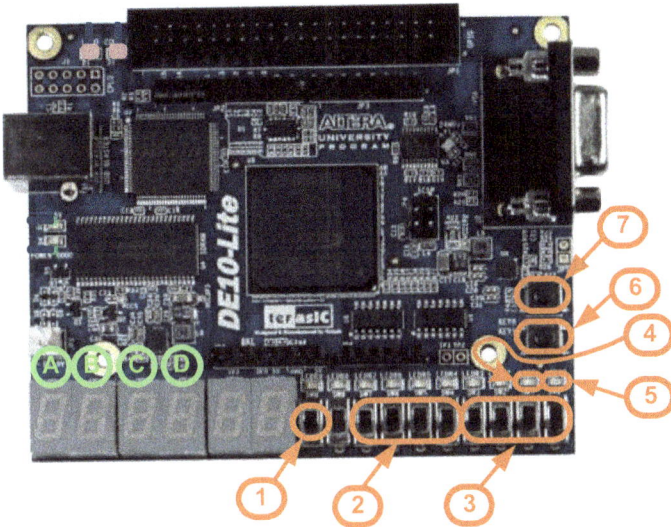

1) the left-most switch, when off (towards the edge of the board) places the board in read mode, while when on (towards the center of the board), places the board in write mode;

2) these four slide switches set the display hex character for either display characters "A" or "C" (as later retrieved from memory), depending which memory is selected;

3) these four slide switches set the display hex character for either display characters "B" or "D" (as later retrieved from memory), depending which memory is selected;

4) when in a write mode, this LED is lit when the entries are being written to the first memory (associated with displays "A" and "C");

5) when in a write mode, this LED is lit when the entries are being written to the second memory (associated with displays "B" and "D"). Note that when in read mode, neither LED is lit;

6) this push-button toggles between the two memories when in write mode, i.e., toggles between the two LEDs;

7) when in write mode, this push-button writes the selected slide-switch nibble value into its associated memory, and increments the memory address. When in read mode, this push-button simply increments the memory address for the next read value.

Memories Exercises

Light another LED when address is zero so that you can track the locations in memory.

Use a slide switch to display the current address.

Automatically and continuously cycle through the 8 addresses.

Chapter Six
Match Game

For this project we'll step up to a fun little game. The user watches a continuing sequence of three character numbers that change randomly every few seconds. The goal is to detect that any two character numbers of one random sample match any two numbers of the next sample. The match can be between two numbers of any order. For example, the following sequence would include these matches:

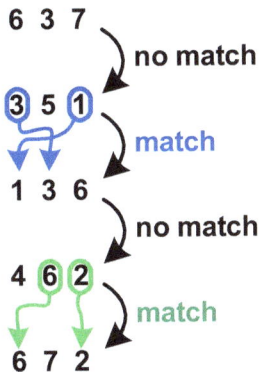

Additionally, we add a bonus combination: duplicated numbers of one sample require just one of the same in the next sample for a match. Like so:

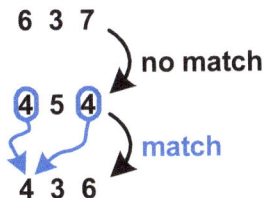

The random numbers span one through eight. A point counter tracks performance, incrementing each time the user successfully tags a match, and decrementing if the user incorrectly tags a non-match or if the user misses a valid match. For our development boards, we use two push-buttons – one to begin a session, and another to tag matches. The design uses four 7-segment display characters – three for the sequence of random number characters, and one for the point tracking.

Here's a first part of the design.

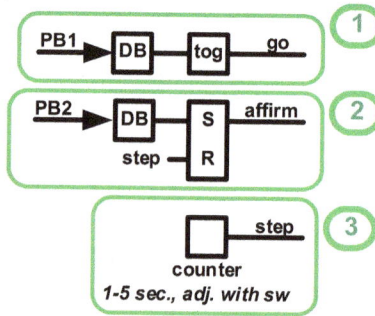

1) the first push-button toggles the "go" signal, which, when set, activates a game session;

2) the second push-button, used to tag a match, sets the "affirm" signal, which is reset at the end of the current random number match cycle period (lasting a few seconds);

3) a free-running counter establishes the random number match cycle periods. A "step" pulse signal flags a transition to the next match period. As we'll see, the cycle period length (i.e., the playing difficulty) can be adjusted using two slide-switches.

Next we look at the operation of the match numbers.

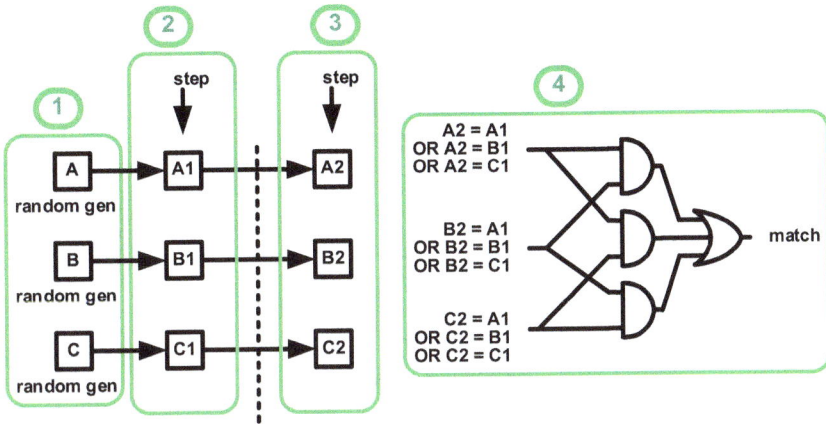

1) we create three random number generators, the sources for the display numbers;

2) at each step time, i.e., the transition from one match cycle period to the next, we latch each of the three the current random numbers in a set of "1" (A1, B1, and C1) registers;

3) and simultaneously we transfer the prior "1" random numbers as new "2" numbers (A2, B2, and C2);

4) during the following cycle period (between this step and the next) we use logic to determine if there are any matches. You can see from the logic that the "match" signal is asserted if any two of the "2" register numbers match any of the "1" numbers. Note that if, for example, A2 and B2 are the same value, then we need only one of that value in the "1" registers to get a match.

To control the game operation, we will use a state machine, a fundamental tool of logic design. As a control mechanism, we define discreet stages of a desired operation that we call states, and the resulting logic state machine passes from one state or stage to another based on conditions. Defined control occurs at each state, and by design, only one state at a time is active.

Here's the state machine for the match game.

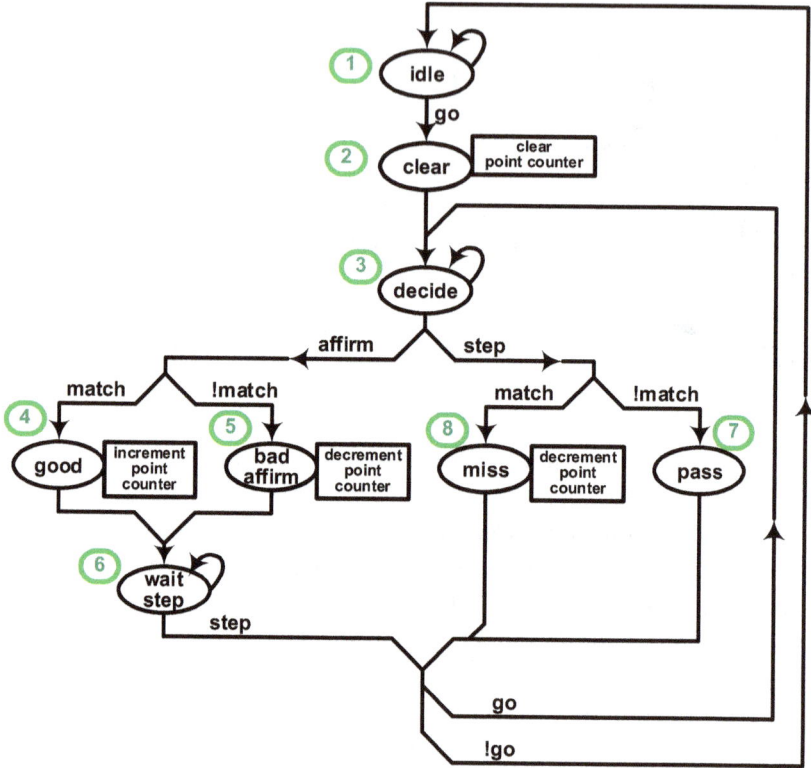

5) if, however, the user guessed wrong, and in fact there is no match this period, then we decrement the point counter and move on;

6) whether the user's guess was correct or not, we now wait patiently for the end of this cycle period, as indicated by the "step" signal, and then, depending on whether the game is still active as indicated by an asserted "go" signal, we return to the "decide" state, ready for the next guess, or, if the "go" signal has been cleared, we return to the "idle" state;

7) from the "decide" state, if the user chooses to not push the affirm button, then at the end of this cycle period, as marked by the "step" pulse, if they guessed correctly that this period does not represent a match, then the state machine transits through the "pass" state, and off to either "decide" or "idle" as already explained – no change to the point counter;

8) if, on the other hand, this period actually did represent a match, that means the user missed it and guessed incorrectly, and we pass through the "miss" state, decrementing the point counter.

Next we dive into the code, and we have a few more new VHDL aspects to introduce.

Here's an outline of the code:

match_game.vhdl

entity declaration

component declarations

signal declarations

package constants

debounce instantiations

affirm and step registers

pseudo-random generators

character translators

match generation

state machine

point counter

seg_drive instantiation

body

architecture

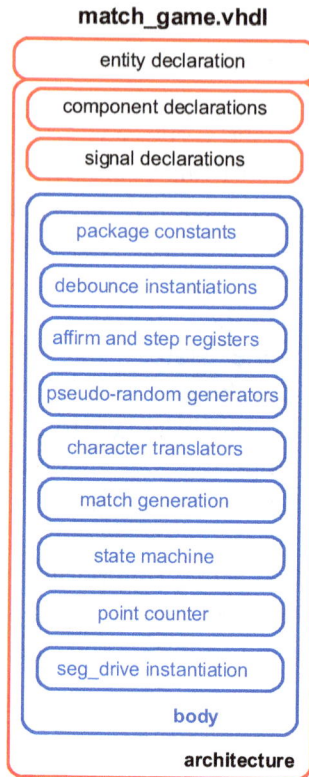

The entity declaration:

```
entity match_game is
  port
   (
   clk            : in  std_logic;
   pb_1           : in  std_logic;
   pb_2           : in  std_logic;
   speed_select : in  std_logic_vector(1 downto 0);
   --
   seven_seg_1  : out std_logic_vector(7 downto 0);
   seven_seg_2  : out std_logic_vector(7 downto 0);
   seven_seg_3  : out std_logic_vector(7 downto 0);
   seven_seg_4  : out std_logic_vector(7 downto 0);
   en1            : out std_logic;
   en2            : out std_logic;
   en3            : out std_logic;
   en4            : out std_logic;
   led_1          : out std_logic;
   led_2          : out std_logic;
   -- unused LEDs
   led          : out  std_logic_vector(9 downto 2)
   );
end entity;
```

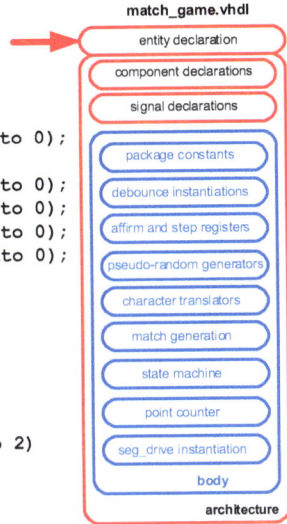

match_game.vhdl

- entity declaration
- component declarations
- signal declarations
- package constants
- debounce instantiations
- affirm and step registers
- pseudo-random generators
- character translators
- match generation
- state machine
- point counter
- seg_drive instantiation

body

architecture

We've seen all these signals before except "speed_select", which is the two left-most slide switches.

Component declarations:

```
-- component declarations
component debounce is
port
  (
  clk           : in    std_logic;
  pb_in         : in    std_logic;
  --
  pb_debounced  : out   std_logic
  );
  end component;

component seg_drive
port
  (
  clk           : in    std_logic;
  mem_dat_1     : in    std_logic_vector(7 downto 0);
  mem_dat_2     : in    std_logic_vector(7 downto 0);
  --
  seven_seg_1   : out std_logic_vector(7 downto 0);
  seven_seg_2   : out std_logic_vector(7 downto 0);
  seven_seg_3   : out std_logic_vector(7 downto 0);
  seven_seg_4   : out std_logic_vector(7 downto 0);
  en1           : out std_logic;
  en2           : out std_logic;
  en3           : out std_logic;
  en4           : out std_logic
  );
  end component;
```

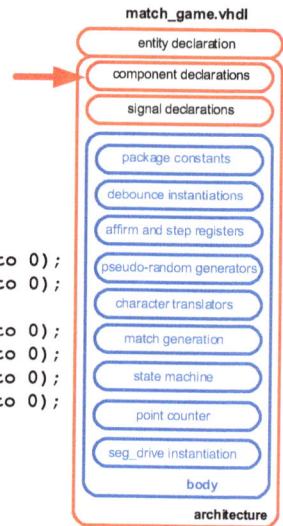

match_game.vhdl

- entity declaration
- component declarations
- signal declarations

- package constants
- debounce instantiations
- affirm and step registers
- pseudo-random generators
- character translators
- match generation
- state machine
- point counter
- seg_drive instantiation

body

architecture

And, again, these are the same as the previous projects.

Next up is the signals declarations, and here we find a new type of signal.

```
type state_type is (
                    idle,
                    clear,
                    decide,
                    good,
                    bad_affirm,
                    miss,
                    pass,
                    wait_step
                    );
signal state : state_type;
```

```
-- signal declarations
signal pb_debncd_1     : std_logic;
signal pb_debncd_1_d1  : std_logic;
signal pb_debncd_2     : std_logic;
signal pb_debncd_2_d1  : std_logic;
signal go              : std_logic;
signal step            : std_logic;
signal affirm          : std_logic;
signal factor1         : std_logic;
signal factor2         : std_logic;
signal factor3         : std_logic;
signal match           : std_logic;
signal step_load       : unsigned(28 downto 0);
signal step_cnt        : unsigned(28 downto 0);
signal prs1            : std_logic_vector(9 downto 1);
signal prs2            : std_logic_vector(9 downto 1);
signal prs3            : std_logic_vector(9 downto 1);
signal next_char_1     : std_logic_vector(3 downto 0);
signal next_char_2     : std_logic_vector(3 downto 0);
signal next_char_3     : std_logic_vector(3 downto 0);
signal match_reg_a1    : std_logic_vector(3 downto 0);
signal match_reg_b1    : std_logic_vector(3 downto 0);
signal match_reg_c1    : std_logic_vector(3 downto 0);
signal match_reg_a2    : std_logic_vector(3 downto 0);
signal match_reg_b2    : std_logic_vector(3 downto 0);
signal match_reg_c2    : std_logic_vector(3 downto 0);
signal point_cnt       : unsigned(3 downto 0);
signal seg_byte_1      : std_logic_vector(7 downto 0);
signal seg_byte_2      : std_logic_vector(7 downto 0);
```

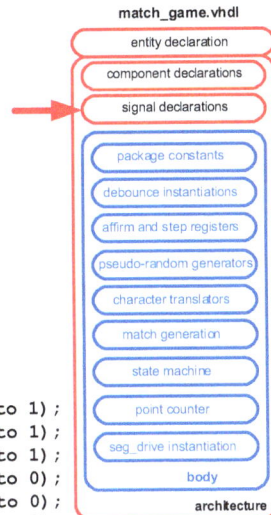

match_game.vhdl

- entity declaration
- component declarations
- signal declarations
- package constants
- debounce instantiations
- affirm and step registers
- pseudo-random generators
- character translators
- match generation
- state machine
- point counter
- seg_drive instantiation

body

architecture

We saw in an earlier project how we could create a new type of signal – in that case, an array for a memory. Here we again create a new signal, one specific for this design, and you may recognize these as the states of the state machine. This is called an enumerated type of signal, and it's just a convenience for clarity. We could have used a regular standard logic signal like this, and it would have functioned the same:

123

```
signal state : std_logic_vector(2 downto 0);
```

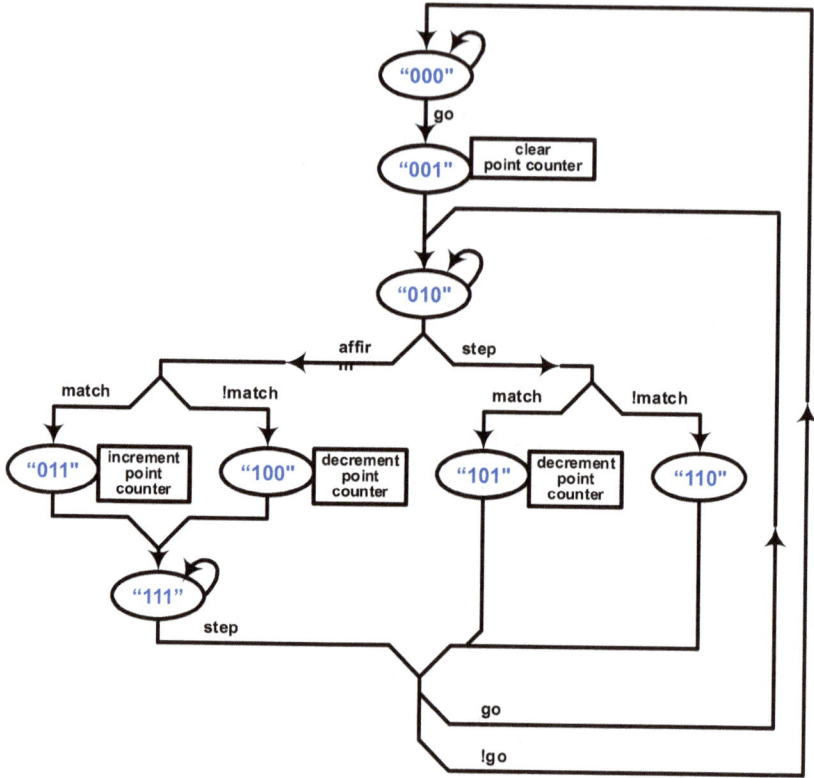

However, by using the same names in the VHDL code as in the state machine diagram, the operation is infinitely easier to understand and work with. Keep in mind that the compiler will translate those names into a set of binary values, similar to our clumsy standard logic version (not necessarily in the same order, of course).

We now move on to the architecture body, and we immediately run into another new VHDL concept.

```
--constants from the package.
step_load <= period1 when speed_select = "00" else
             period2 when speed_select = "01" else
             period3 when speed_select = "10" else
             period4;                    --"11"
```

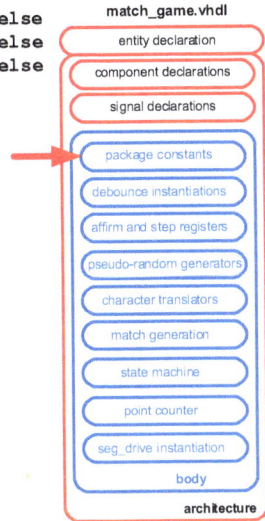

match_game.vhdl

entity declaration

component declarations

signal declarations

package constants

debounce instantiations

affirm and step registers

pseudo-random generators

character translators

match generation

state machine

point counter

seg_drive instantiation

body

architecture

Notice that here we're referring to signals "period1" through "period4" that were not included in the signal declarations above. That's because these are constants that are declared in a separate file, called a "package." A package can be thought of as a place where common pieces of VHDL code are stored, and from where different files can access them. We have to tell each file that intends to use the package's contents about it, though. This is done up at the top with the library declarations, since a package serves more or less the same purpose.

```
-- Project: match_game
--
-- game to match sequences of numbers.

library IEEE;
use IEEE.STD_LOGIC_1164.all;
use IEEE.STD_LOGIC_MISC.all;
use IEEE.STD_LOGIC_UNSIGNED.all;
use IEEE.NUMERIC_STD.all;

use work.projects_pkg.all;

entity match_game is
  port
    (
    clk           : in  std_logic;
    pb_1          : in  std_logic;
    pb_2          : in  std_logic;
```

In this case, we're naming our package file "project_pkg", although it can be anything.

There's a reason we're using a package for these constants; just as we used different "seg_control" and "seg_drive" modules for the Terasic and Digilent boards, we need different constants here for the two boards. And, just as we were able to use one common module instantiation for the two board projects, here we're able to use one "step_load" signal assignment for both boards. In both cases, we have different files for the two boards ("seg_control", "seg_drive", and now also "project_pkg"), and we point to the appropriate one within the project file assignments.

Here's the entire package contents for the Terasic board.

```
library IEEE;
use IEEE.STD_LOGIC_1164.all;
use IEEE.STD_LOGIC_MISC.all;
use IEEE.STD_LOGIC_UNSIGNED.all;
use IEEE.NUMERIC_STD.all;

-- For the Terasic board -- 50MHz clock

package projects_pkg is                          1

  constant period1  : unsigned(28 downto 0) := '1' & X"0000000";
  constant period2  : unsigned(28 downto 0) := '0' & X"A800000";
  constant period3  : unsigned(28 downto 0) := '0' & X"6000000";
  constant period4  : unsigned(28 downto 0) := '0' & X"3000000";

end package projects_pkg;

package body projects_pkg is                     2

end package body projects_pkg;
```

1) we have the necessary keywords "package" and "is", and the name of this package, "projects_pkg", which, like all our design files, is the same as the file name ("projects_pkg.vhdl"). Like file entity declarations, we also declare the end of this declaration section of the package with "end package";

2) our simple package has no body (just constant declarations), but VHDL requires that we include the whole structure anyway.

The purpose of this package is to establish four constants.

```
library IEEE;
use IEEE.STD_LOGIC_1164.all;
use IEEE.STD_LOGIC_MISC.all;
use IEEE.STD_LOGIC_UNSIGNED.all;
use IEEE.NUMERIC_STD.all;

-- For the Terasic board -- 50MHz clock

package projects_pkg is

    constant period1 : unsigned(28 downto 0) := '1' & X"0000000";
    constant period2 : unsigned(28 downto 0) := '0' & X"A800000";
    constant period3 : unsigned(28 downto 0) := '0' & X"6000000";
    constant period4 : unsigned(28 downto 0) := '0' & X"3000000";

end package projects_pkg;

package body projects_pkg is

end package body projects_pkg;
```

```
-- period1 = ~5   seconds
-- period2 = ~3.5 seconds
-- period3 = ~2   seconds
-- period4 = ~1   second
```

1) constants are declared using the (obvious) keyword "constant",

2) followed by the name of the constant – note that these are the names used in the "match_game" file above;

3) after we declare the type (in this case unsigned, since that's the type of signal these will be assigned to), a ":=" flags that the constant value follows;

4) at the 50MHz clock of the Terasic board, these countdown values,

5) yield these commented times.

The "&" indicates concatenation (gluing the two parts together):

```
constant period1 : unsigned(28 downto 0) := '1' & X"0000000" ;
```

```
"10000000000000000000000000000"
```

So, the constant value for "period1" is a binary one followed by 28 binary zeros.

This is the package for the Digilent board.

```
library IEEE;
use IEEE.STD_LOGIC_1164.all;
use IEEE.STD_LOGIC_MISC.all;
use IEEE.STD_LOGIC_UNSIGNED.all;
use IEEE.NUMERIC_STD.all;

-- For the Digilent board -- 100MHz clock

package projects_pkg is

   constant period1  : unsigned(28 downto 0) := '1' & X"FFFFFFF";
   constant period2  : unsigned(28 downto 0) := '1' & X"5000000";
   constant period3  : unsigned(28 downto 0) := '0' & X"C000000";
   constant period4  : unsigned(28 downto 0) := '0' & X"6000000";

end package projects_pkg;

package body projects_pkg is

end package body projects_pkg;
```

They are the same, except that since the Digilent board's clock is 100MHz, the countdown values are twice as large to get the same time periods. Note that:
('1' & X"FFFFFFF") is effectively the same as ('2' & "0000000").

Next come the debounce instantiations,

```
debounce_1 : debounce
port map
  (
   clk          => clk,       --in   std_logic;
   pb_in        => pb_1,      --in   std_logic;
   --
   pb_debounced => pb_debncd_1 --out  std_logic
  ) ;

debounce_2 : debounce
port map
  (
   clk          => clk,       --in   std_logic;
   pb_in        => pb_2,      --in   std_logic;
   --
   pb_debounced => pb_debncd_2 --out  std_logic
  ) ;
```

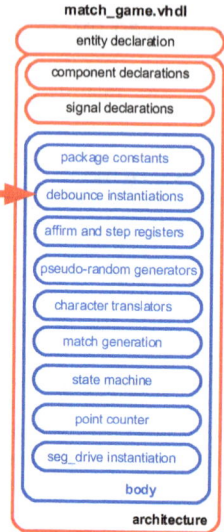

match_game.vhdl

- entity declaration
- component declarations
- signal declarations
- package constants
- debounce instantiations
- affirm and step registers
- pseudo-random generators
- character translators
- match generation
- state machine
- point counter
- seg_drive instantiation
- body

architecture

and these are the same as we've seen before.

Next up, we have registers that implement the "go", "affirm", and "step" signals.

```
debounce_proc : process(clk)
begin
  if rising_edge(clk) then
    pb_debncd_1_d1 <= pb_debncd_1;
    if (pb_debncd_1 = '0' and pb_debncd_1_d1 = '1') then
      go <= not go;
    end if;
    --
    pb_debncd_2_d1 <= pb_debncd_2;
    if (pb_debncd_2 = '0' and pb_debncd_2_d1 = '1') then
      affirm <= '1';
    elsif (step = '1') then
      affirm <= '0';
    end if;
  end if;
end process;

timer : process(clk)
begin
  if rising_edge(clk) then
    if (step_cnt = '0' & X"0000000") then
      step_cnt <= step_load;
      step <= '1';
    else
      step_cnt <= step_cnt - 1;
      step <= '0';
    end if;
  end if;
end process;

led_1 <= go;
led_2 <= affirm;
```

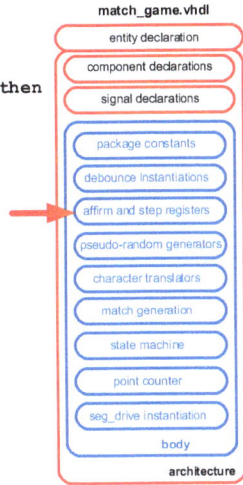

as we saw from our initial block diagram.

```
debounce_proc : process(clk)
begin
   if rising_edge(clk) then
      pb_debncd_1_d1 <= pb_debncd_1;
      if (pb_debncd_1 = '0' and pb_debncd_1_d1 = '1') then
         go <= not go;
      end if;
      --
      pb_debncd_2_d1 <= pb_debncd_2;
      if (pb_debncd_2 = '0' and pb_debncd_2_d1 = '1') then
         affirm <= '1';
      elsif (step = '1') then
         affirm <= '0';
      end if;
   end if;
end process;

timer : process(clk)
begin
   if rising_edge(clk) then
      if (step_cnt = '0' & X"0000000") then
         step_cnt <= step_load;
         step <= '1';
      else
         step_cnt <= step_cnt - 1;
         step <= '0';
      end if;
   end if;
end process;

led_1 <= go;
led_2 <= affirm;
```

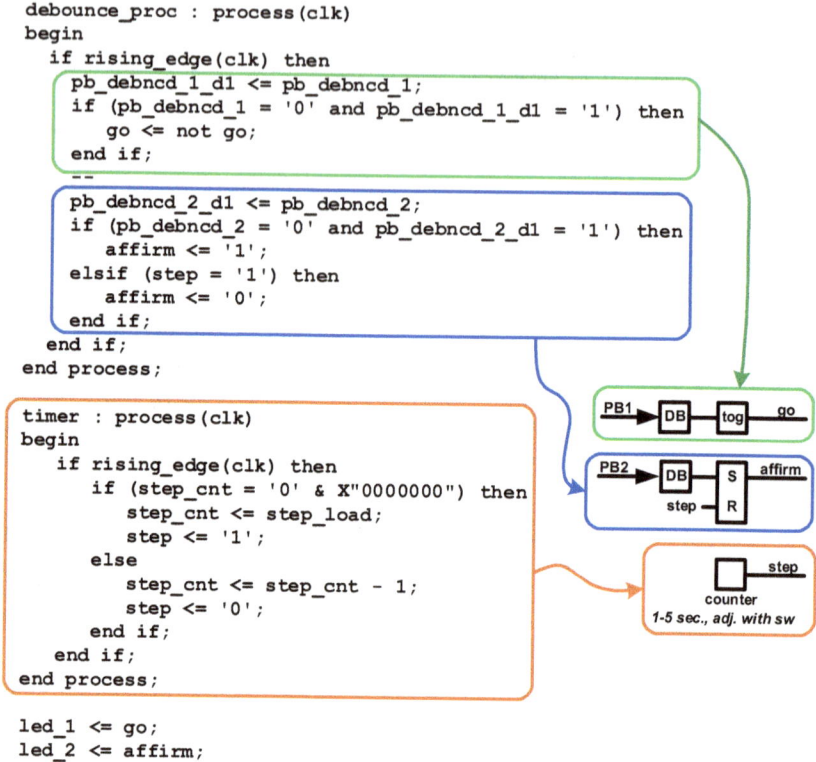

A bit of explanation is in order from the step signal generation.

```
timer : process(clk)
begin
   if rising_edge(clk) then           (1)
      if (step_cnt = '0' & X"0000000") then
         step_cnt <= step_load;
         step <= '1';
      else
         step_cnt <= step_cnt - 1;      (2)
         step <= '0';
      end if;
   end if;
end process;
```

1) the counter ("step_cnt") is continually decrementing. As it passes through zero, we reload it with the constant from above ("step_load"), and assert the "step" signal;

2) on the very next clock, we de-assert the "step" signal, and the counter begins counting down again.

And now it's time to take another look at those constants.

```
--constants from the package.
step_load <= period1 when speed_select = "00" else
             period2 when speed_select = "01" else
             period3 when speed_select = "10" else
             period4;                          --"11"



timer : process(clk)
begin
    if rising_edge(clk) then
        if (step_cnt = '0' & X"0000000") then
            step_cnt <= step_load;
            step <= '1';
        else
            step_cnt <= step_cnt - 1;
            step <= '0';
        end if;
    end if;
end process;
```

1) the down-counter load value "step_load",
2) is selected from among the package constants based on "speed_select", which is a combination of two of the slide switches (the two right-most).

You may have been wondering how we're able to get random numbers from VHDL code, and we're about to find out. But first, we need to note that, as the label for this next section indicates, we will not be creating strictly random numbers, but what we call pseudo-random values, meaning that, although the values from sample to sample are effectively random, after some amount of time, the sequence repeats. Given enough time, a devoted player might begin to memorize the combinations of number sequences, but, as we'll see, we add an additional twist by loading each of the three pseudo-random generators with different initial "seed" values every run, thus ensuring that each game is different.

Blaine C. Readler

The mathematical theory behind pseudo-random generation is somewhat involved, and if you're interested, Wikipedia generally has very good coverage. For our purposes, though, we'll look at a basic implementation, very common across many designs. The idea is to use a shift register with feedback – the output back to the input – to endlessly cycle, but where we mix up the stream by XORing one or more points, called taps, along the way with the fed-back output. The number of points and locations of XOR'd insertion in the shifted stream determine how long the cycle goes before repeating. Here's ours.

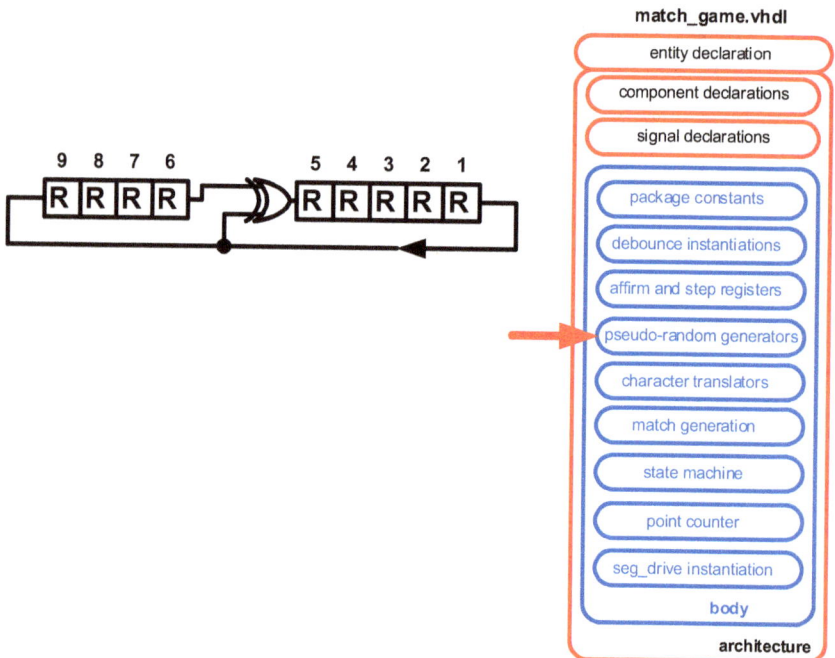

The first thing we note is that our shift register has nine stages (registers). This allows up to 2^9, i.e. 512, possible combinations (actually only 511, since all zeros would lock out the operation). This is how many pseudo-random numbers we would get before the pattern starts over. However, not every combination of XOR'd feedback will give all possible combinations. Those configurations that do are referred to as maximal arrangements, and ours is one of those.

Still avoiding in-depth theory, you should know that this type of pseudo-random generation is referred to as a Linear-Feedback Shift Register, or LFSR. Again, Wikipedia should be a good source. We define pseudo-random generators using a polynomial, and this is ours.

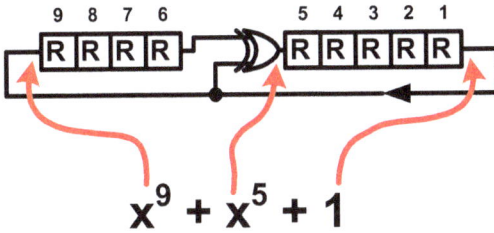

$$x^9 + x^5 + 1$$

Here's the code that implements this polynomial pseudo-random generator.

```
                  (1)
signal prs1   : std_logic_vector(9 downto 1);
signal prs2   : std_logic_vector(9 downto 1);
signal prs3   : std_logic_vector(9 downto 1);

prs_process : process(clk)
begin                                          (2)
   if rising edge(clk) then
      if (state = clear) then --load seeds
         prs1 <= std_logic_vector(step_cnt(8 downto 0));
         prs2 <= std_logic_vector(step_cnt(17 downto 9));
         prs3 <= std_logic_vector(step_cnt(26 downto 18));
      else
         prs1 <=    prs1(1)
                 & prs1(9 downto 7)
                 & (prs1(6) xor prs1(1))
                 & prs1(5 downto 2);              (3)

         prs2 <=    prs2(1)
                 & prs2(9 downto 7)
                 & (prs2(6) xor prs2(1))
                 & prs2(5 downto 2);

         prs3 <=    prs3(1)
                 & prs3(9 downto 7)
                 & (prs3(6) xor prs3(1))
                 & prs3(5 downto 2);
      end if;
   end if;
end process;
```

1) this is just a recap, showing the signal declaration for the pseudo-random shift register. Note that instead of "(8 downto 0)", we're using "(9 downto 1)". Either works, but the latter makes it easier to correlate the theory with the code;

2) at the beginning of each game run we load the shift registers with different portions of the free-running step counter. This is the random "seed," which serves to effectively give each of the three generators a different starting point for every run;

3) this is one of the three pseudo-random generator shift registers, and we'll look at this separately.

We begin by looking at two different assignments.

```
prs1(9 downto 1) <= prs1(9 downto 1);     ①

prs1(9 downto 1) <= prs1(1) & prs1(9 downto 2);    ②
```

1) you can see that this is not a shift register, since with each clock, each bit is returned to the same position;

2) this one is a shift register.

This is how the shift register works.

```
                    ①              ②
prs1(9 downto 1) <= prs1(1) & prs1(9 downto 2);
```

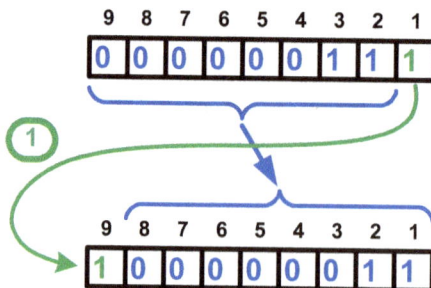

1) each clock, the LS bit is copied to the MS position,

2) and the rest of the MS part is copied one position to the right.

You can see that with each clock, the LS bit is shifted around to the MS, and the rest are shifted to the right.

Finally, we'll introduce the XOR'd tap, which transforms the shift register into a pseudo-random number generator, and we'll take it in two steps.

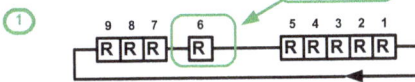

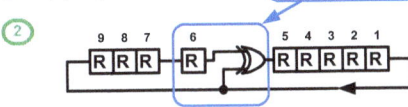

1) first, we simply break the shift register we just created into parts;

2) and then we add the XOR gating.

We now have a nine-stage pseudo-random generator, creating a 9-bit number from which we need to create a series of 7-segment seemingly random characters. The first question we ask is how many different characters for each display position we'd like in the game. For example, we could randomly display sixteen characters -- everything from "0" to hex "F." This wouldn't work very well for our game operation, however, since the chance of any two characters being repeated between steps is small, and the user would have to wait too long for a match to occur.

On the other hand, too few characters – say just four – would produce far too many matches, and our point counter (just one display from 0-9) would max out almost immediately.

I've chosen eight characters as the Goldilocks "just right" number. Not only does this provide a satisfying amount of matches, but, as it happens, eight is a convenient logic value.

We must translate three pseudo-random outputs to three "next_char" values to be used for our A1/B1/C1 match registers.

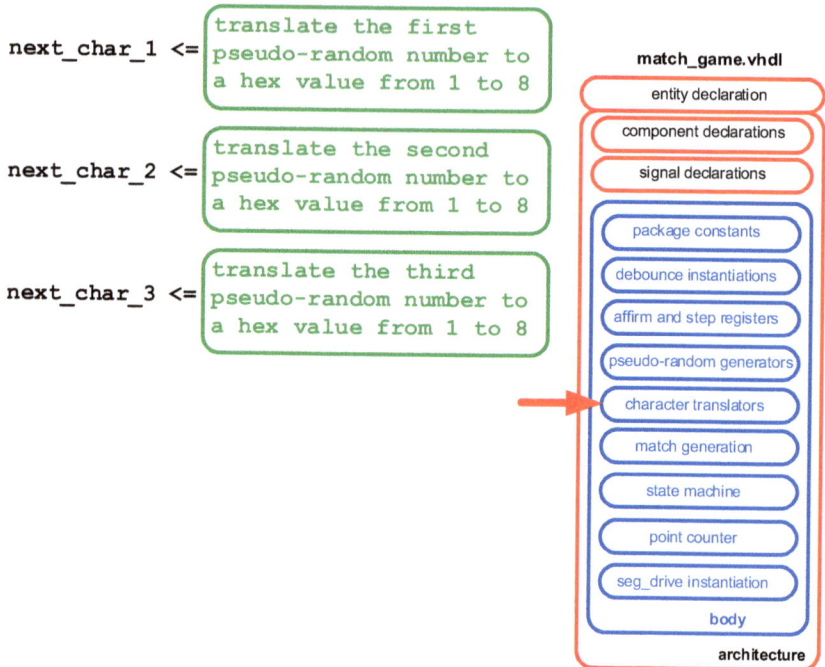

```
next_char_1 <=  translate the first
                pseudo-random number to
                a hex value from 1 to 8

next_char_2 <=  translate the second
                pseudo-random number to
                a hex value from 1 to 8

next_char_3 <=  translate the third
                pseudo-random number to
                a hex value from 1 to 8
```

match_game.vhdl

- entity declaration
- component declarations
- signal declarations
  - package constants
  - debounce instantiations
  - affirm and step registers
  - pseudo-random generators
  - → character translators
  - match generation
  - state machine
  - point counter
  - seg_drive instantiation
  - **body**

architecture

And we'll look at the first translation:

next_char_1 <= translate the first pseudo-random number to a hex value from 1 to 8

prs1[9]

prs1[8]

next_char_1

prs1[7]

x"0" } 1
x"1"

prs1[6]

x"2" } 2
x"3"

prs1[5]

x"4" } 3
x"5"

prs1[4]

x"6" } 4
x"7"

prs1[3]

x"8" } 5
x"9"

prs1[2]

x"A" } 6
x"B"

prs1[1]

x"C" } 7
x"D"

①

x"E" } 8
x"F"

②

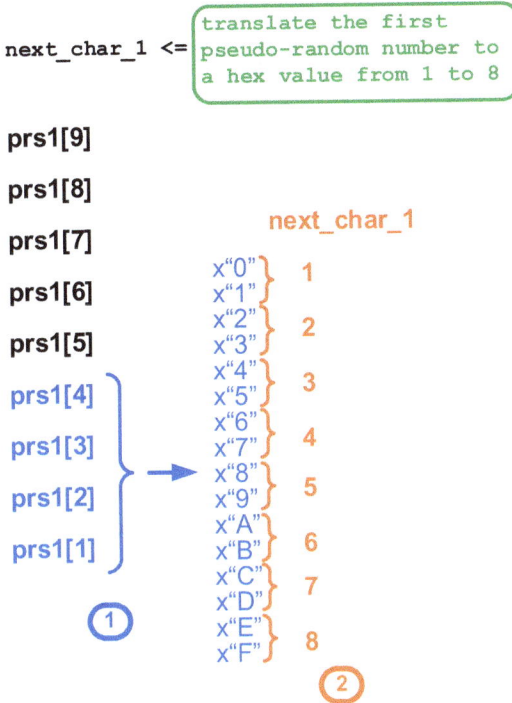

1) of the nine bits of the pseudo-random value, we use just four (noting that any portion of a random number has the same degree of randomness). We then treat these four bits as the four bits of a hex number;

2) we then "down-sample" the resulting possible sixteen values to the eight that we're after, which we assign to "next_char_1".

Here's the VHDL code.

```
                                    (1)
next_char_1 <= X"1" when     (prs1(4 downto 1) >= X"0")
                         and (prs1(4 downto 1) <= X"1") else
              X"2" when      (prs1(4 downto 1) >= X"2")        (2)
                         and (prs1(4 downto 1) <= X"3") else
              X"3" when      (prs1(4 downto 1) >= X"4")
                         and (prs1(4 downto 1) <= X"5") else
              X"4" when      (prs1(4 downto 1) >= X"6")
                         and (prs1(4 downto 1) <= X"7") else
              X"5" when      (prs1(4 downto 1) >= X"8")
                         and (prs1(4 downto 1) <= X"9") else
              X"6" when      (prs1(4 downto 1) >= X"A")
                         and (prs1(4 downto 1) <= X"B") else
              X"7" when      (prs1(4 downto 1) >= X"C")
                         and (prs1(4 downto 1) <= X"D") else
              X"8";
```

1) using just four bits of the pseudo-random value;

2) "down-sampling" the resulting possible sixteen values to eight, which we assign to "next_char_1"   The ">=" combination here means "greater-than-or-equal-to", and "<=" means "less-than-or-equal-to".

Up next is the match generation, where we determine if the displayed characters have resulted in a match.

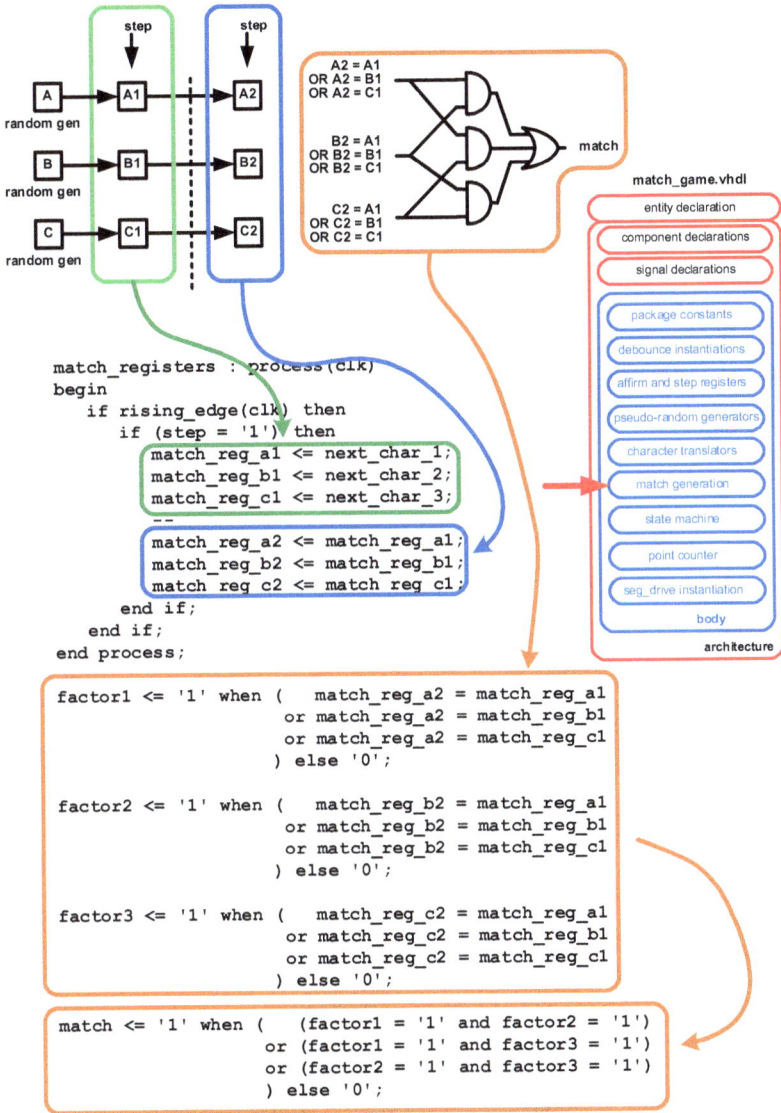

```
match_registers : process(clk)
begin
    if rising_edge(clk) then
        if (step = '1') then
            match_reg_a1 <= next_char_1;
            match_reg_b1 <= next_char_2;
            match_reg_c1 <= next_char_3;
            --
            match_reg_a2 <= match_reg_a1;
            match_reg_b2 <= match_reg_b1;
            match_reg_c2 <= match_reg_c1;
        end if;
    end if;
end process;
```

```
factor1 <= '1' when (   match_reg_a2 = match_reg_a1
                     or match_reg_a2 = match_reg_b1
                     or match_reg_a2 = match_reg_c1
                    ) else '0';

factor2 <= '1' when (   match_reg_b2 = match_reg_a1
                     or match_reg_b2 = match_reg_b1
                     or match_reg_b2 = match_reg_c1
                    ) else '0';

factor3 <= '1' when (   match_reg_c2 = match_reg_a1
                     or match_reg_c2 = match_reg_b1
                     or match_reg_c2 = match_reg_c1
                    ) else '0';
```

```
match <= '1' when (   (factor1 = '1' and factor2 = '1')
                   or (factor1 = '1' and factor3 = '1')
                   or (factor2 = '1' and factor3 = '1')
                  ) else '0';
```

We come to the state machine:

Blaine C. Readler

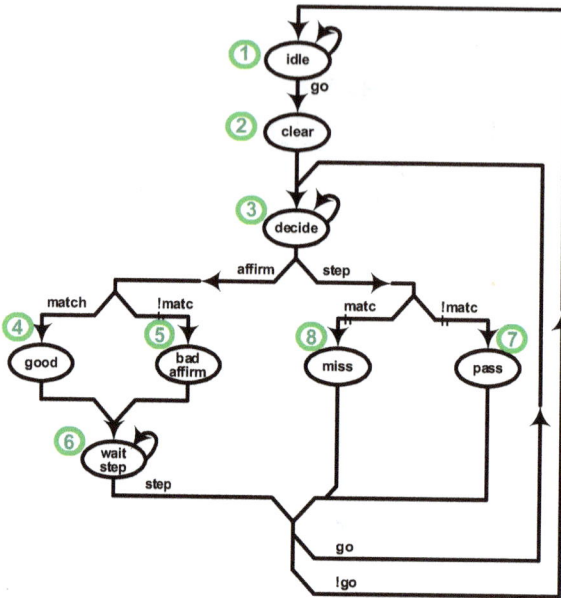

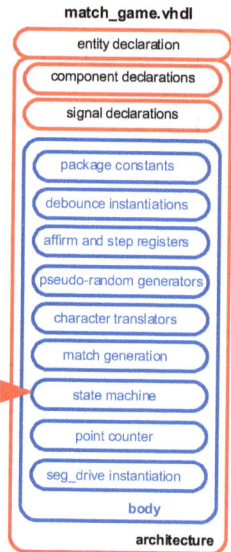

```
state_machine : process(clk)
begin
  if rising_edge(clk) then
    case (state) is
   1 when idle     => if (go = '1') then        state <= clear;  end if;
   2 when clear    =>                            state <= decide;
   3 when decide   => if (affirm = '1') then
                        if (match = '1') then state <= good;
                        else                  state <= bad_affirm; end if;
                      elsif (step = '1') then
                        if (match = '1') then state <= miss;
                        else                  state <= pass; end if;
                      end if;
   4 when good      =>                           state <= wait_step;
   5 when bad_affirm =>                          state <= wait_step;
   6 when wait_step => if (step = '1') then
                        if (go = '1') then  state <= decide;
                        else                state <= idle; end if;
                      end if;
   8 when miss      => if (go = '1') then   state <= decide;
                      else                  state <= idle; end if;
   7 when pass      => if (go = '1') then   state <= decide;
                      else                  state <= idle; end if;
      when others   => state <= idle;
    end case;
  end if;
end process;
```

We use a case statement, where the case selection is made using the enumerated labels we assigned to the "state" signal. By definition, the "state" signal can only be one label at a time. The location within the state machine at any time is represented by a

142

"when" entry in the case statement. You can see that the only action taken in each "when" entry is to assig a different state label to "state". For example, when we're in the "idle" state, we stay there until the "go" signal goes active, at which point (the next clock) the "state" signal goes to "clear", and we're now at that "when" entry. From there we go directly to "decide" with no condition, just like in the state machine diagram.

Note that for conditional states (where we wait for something to happen to move on) we don't need to explicitly tell the case entry to stay there if the condition is not met, since that's the case default. For example, when in the "idle" state, we do not need to include:

"elsif (go ='0') then state <= idle".

Here's the next operation, the point counter.

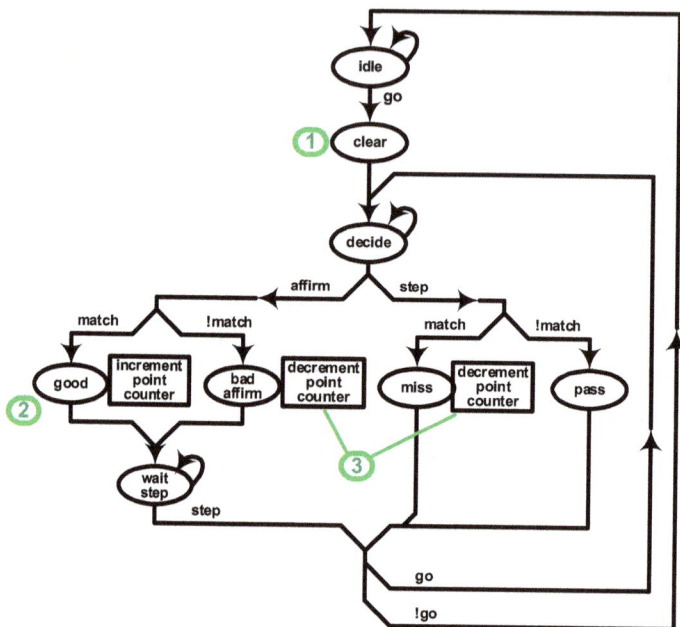

```
point_counter : process(clk)
begin
   if rising_edge(clk) then
①   if (state = clear) then
        point_cnt  <= X"0";
②   elsif (   point_cnt /= X"9"
           and state = good
        ) then
        point_cnt  <= point_cnt + 1;
③   elsif (   point_cnt /= 0
           and (   state = bad_affirm
               or state = miss
               )
        ) then
        point_cnt  <= point_cnt - 1;
     end if;
   end if;
end process;
```

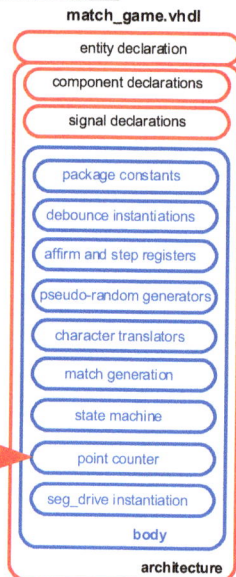

match_game.vhdl

entity declaration

component declarations

signal declarations

package constants

debounce instantiations

affirm and step registers

pseudo-random generators

character translators

match generation

state machine

point counter

seg_drive instantiation

body

architecture

1) each time the user starts a new game, the clear state clears the point counter;

2) the point counter is incremented when the user correctly tags a match. Note that once the counter reaches "9", it

doesn't increment any more – this marks a successful game run. Continuing past this would require that the user be familiar with hex characters;

3) the counter is decremented with either an incorrectly tagged match (bad_affirm) or a legitimate match that the user failed to tag (miss).

The last part of the architecture body is the "seg_drive" module instantiation. We saw this instantiated in the previous project, and here as well there are separate versions for the two project development boards. There are, however, a few specifics to look at.

```vhdl
seg_byte_1 <= match_reg_a1 & match_reg_b1 when go = '1'    1
                                          else X"00";
                                                               2
seg_byte_2 <= match_reg_c1 & std_logic_vector(point_cnt) when go = '1'
                                          else X"00";

seg_drive_i : seg_drive
port map
  (
  clk        4    => clk,      3    --in    std_logic;
  mem_dat_1       => seg_byte_1,    --in    std_logic_vector(7 downto 0);
  mem_dat_2       => seg_byte_2,    --in    std_logic_vector(7 downto 0);
  --
  seven_seg_1  => seven_seg_1,  --out   std_logic_vector(7 downto 0);
  seven_seg_2  => seven_seg_2,  --out   std_logic_vector(7 downto 0);
  seven_seg_3  => seven_seg_3,  --out   std_logic_vector(7 downto 0);
  seven_seg_4  => seven_seg_4,  --out   std_logic_vector(7 downto 0);
  en1          => en1,          --out   std_logic;
  en2          => en2,          --out   std_logic;
  en3          => en3,          --out   std_logic;
  en4          => en4           --out   std_logic
  );
```

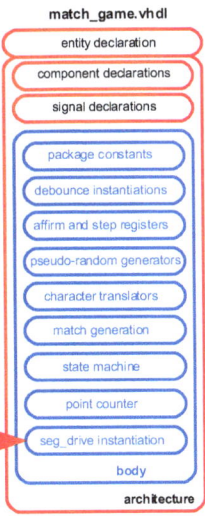

145

1) each "seg_byte" drives two 7-segment display characters, so we concatenate (with "&") the first two four-bit match registers together to display the first two random characters of the game. If "go" is not asserted, meaning that the game is currently idle, then we display zeros, which, remember, are not included in the game's random collection of possible characters;

2) the last two 7-segment display characters are comprised of the third random character and, lastly, the point counter;

3) the concatenated collection of display characters are presented to the instantiated "seg_drive" module,

4) where we still have the signal names from the previous project ("memories"). We could have created two whole new versions of "seg_byte" specifically for this project, but since module reuse is commonplace, here's some practice using it.

Terasic/Quartus

A) these 7-segment displays comprise the three match game random characters;

B) the point counter; stops at a value of 9;

1) this push-button toggles the game off and on. Toggling the game off clears the displays, including the point counter;

2) the active-game match tag, pushed any time a match is believed to have occurred;

3) LED indicating that a game is in progress (when lit);

4) LED indicating that a match tag has been declared – stays lit for the duration of the current cycle period;

5) game cycle period setting. Four possible settings:

"00" = ~5 seconds
"01" = ~3.5 seconds
"10" = ~2 seconds
"11" = ~1 second

Digilent/Vivado

A) these 7-segment displays comprise the three match game random characters;

B) the point counter; stops at a value of 9;

1) this push-button toggles the game off and on. Toggling the game off clears the displays, including the point counter;

2) the active-game match tag, pushed any time a match is believed to have occurred;

3) LED indicating that a game is in progress (when lit);

4) LED indicating that a match tag has been declared – stays lit for the duration of the current cycle period;

5) game cycle period setting. Four possible settings:
"00" = ~5 seconds
"01" = ~3.5 seconds
"10" = ~2 seconds
"11" = ~1 second

Match Game Exercises

Currently, a game run continues regardless of the point count. Declare a point count of 9 as the successful end of the game, perhaps blinking the point count LED in celebration.

Ensure that the pseudo-random generator doesn't load with zero (which would freeze it).

Chapter Seven
FIFO

In this project we introduce a FIFO memory (First-In, First-Out), and we continue in the game mode; in this case using the individual segments of the 7-segment displays rather than full numerical characters. Here, the user must remember the position of a tagged segment among a sequence presented on the first display, and then tag when that same segment appears among the sequence of the second display, while at the same time controlling the flow of the sequence through the FIFO.

Here we see an example of a sequence of display values, where each row represents the outputs within a succession of cycle steps:

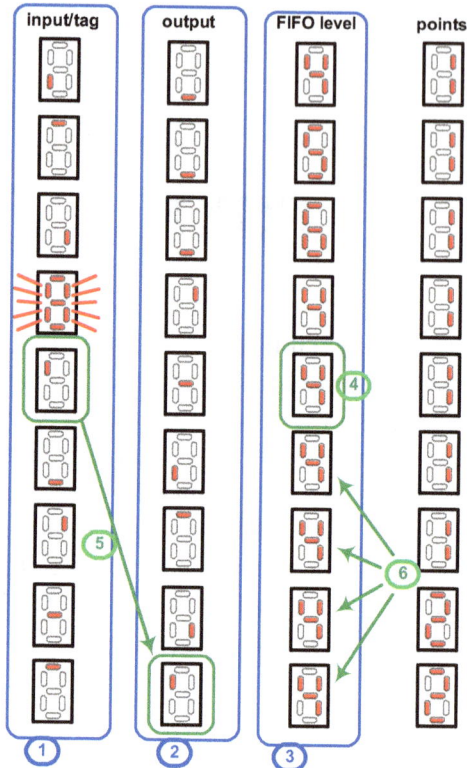

1) these segment locations are randomly created and written into the FIFO input. The user observes these, waiting to be shown which one to remember;

2) these are the segments as read from the FIFO, previously written. The user has direct control over the reads made to the FIFO;

3) this display shows the user the current FIFO level, i.e., how many entries are still in the FIFO. Another way to look at this is how many FIFO reads must be done to see the segment currently shown in the first display;

4) for example, since the FIFO level in this example is "4",

5) the upper-left segment appears four cycle steps later (assuming the user continued reading the FIFO);

6) here, the user continually read from the FIFO each cycle step, and so the FIFO level remains at "4".

The user uses one of the push-buttons, referred to as "next," to read the FIFO. The game (i.e., the VHDL code) never lets the FIFO fall below four entries. Thus, the user may have pushed the "next" button multiple times between cycle steps, but the FIFO would always remain at four entries.

For example,

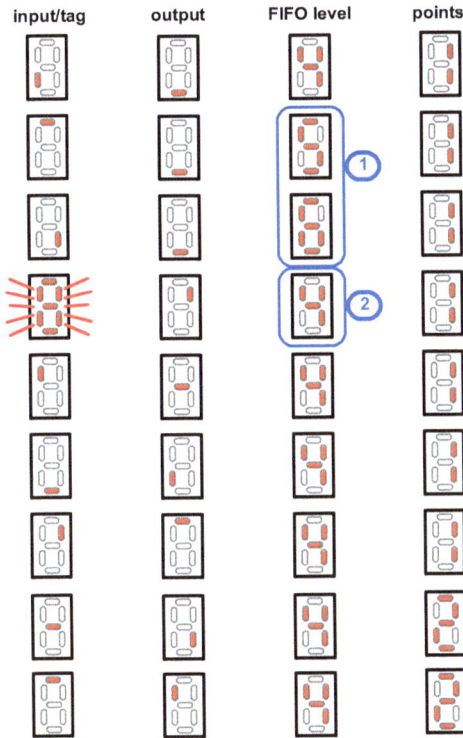

1) here the user chose to not read the FIFO for two consecutive cycle steps, and so the FIFO level rose from 4 to 6;

2) but during the next cycle-step, the user pushed the "next" button three times to get back to 4 (noting that the level rose to 7 before the user began pushing the "next" button).

Let's see how a point is gained.

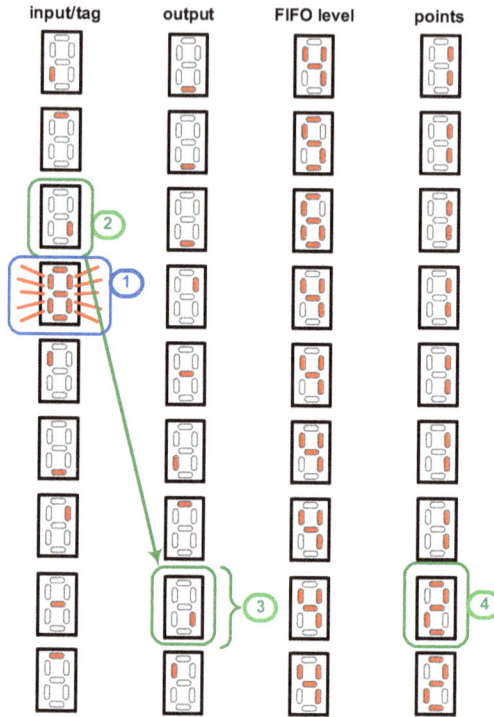

1) at random times, the FIFO input display will flash. This is the "tag" indication;

2) the user will need to have noted the segment location PRIOR to the tag flash (i.e., they must note and remember in turn each segment in the sequence);

3) the user then waits until that segment location appears at the FIFO output as the user continues to read the FIFO (using the "next" push-button);

4) if the user correctly pushes the "affirm" push-button during this cycle-step, then

5) they achieve a point, and the point display increments.

Here, the user missed the proper cycle step, and . . .

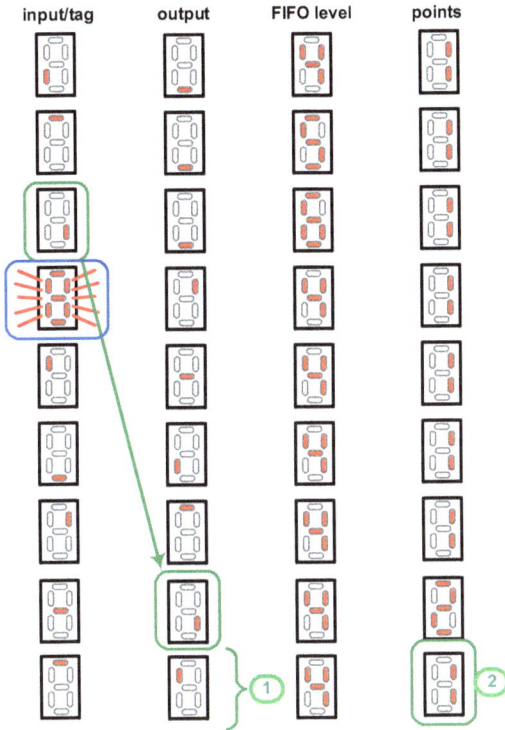

1) pushed the "affirm" button too late,
2) and so lost one point.

If the user allows the 16-word FIFO to fill (all the way to hex 0xF), the game ends, with both the input and output displays continually flashing.

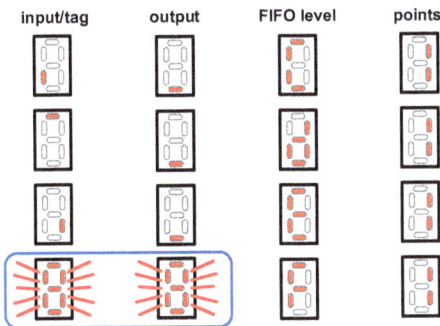

A new game is begun by holding both the "next" and "affirm" push-buttons low for a few seconds.

Here's the block diagram, showing three new modules:

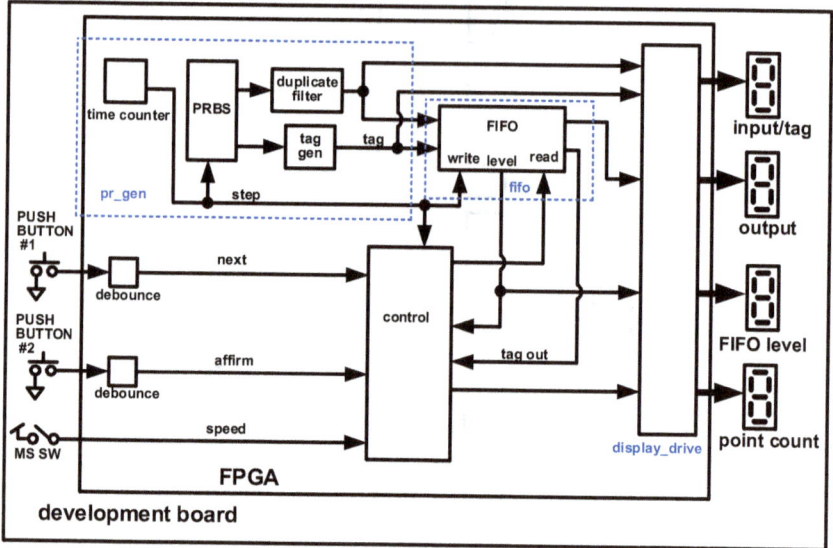

"pr_gen" – generates a repeating "step" indication every couple of seconds, a pseudo-random generator (PRBS – Pseudo-Random Binary Sequence), a random "tag" generator, and a filter to suppress duplicate consecutive segments;

"fifo" – as the name indicates, this is the first-in-first-out memory; and

"display_drive", which creates 7-segment characters for the last two displays as in previous projects, but also creates the individual line segments unique to this game for the first two displays.

The core of the control block shown in the diagram is a state machine:

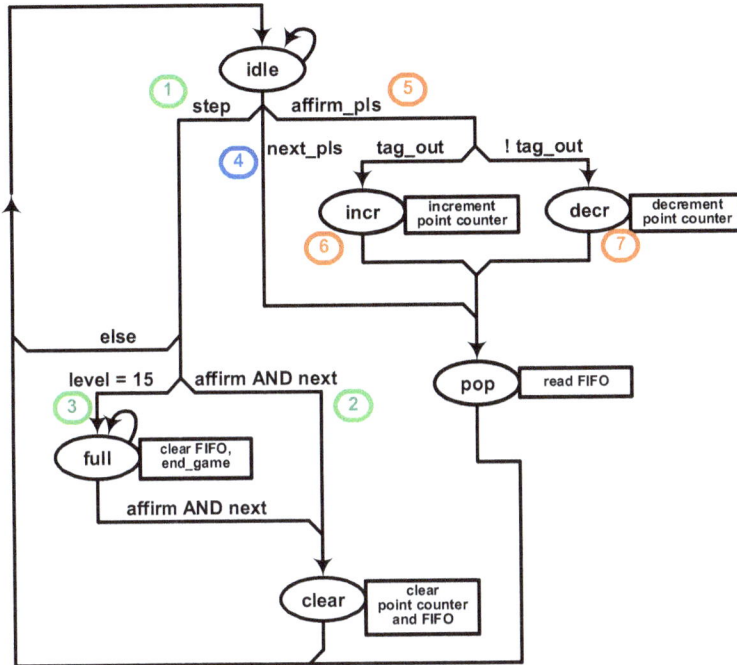

1) we exit the "idle" state at the end of a step cycle period if the user hasn't pushed just one button since the last step;

2) if the user is holding both buttons down, however, we clear the point counter and FIFO, and return to the "idle" state to begin a new game;

3) the game ends if the FIFO level has reached its max value, 0xF (15), in which case we clear the FIFO and flash the two displays continuously until, again, the user holds both buttons down to begin a new game.

4) if during a step period the user pushes the "next" button, we read the next word from the FIFO – called a "pop" – and return to the "idle" state;

5) if, instead, the user pushes the "affirm" button, indicating that they think that a tag word has come through the FIFO, then either,

6) the entry read from the FIFO is either indeed a tag (indicated by a "tag_out" bit, which was carried through the

FIFO along with the tag segment), in which case the point counter is incremented, or

7) the entry read from the FIFO is not a tag, and so the point counter is decremented. In either case, the FIFO is popped and we return to the "idle" state.

We now move on to the VHDL code. This game is more involved than the previous match game, and, as you'll see, we become acquainted with a greater degree of coding complexity.

Here's the layout of the top VHDL file. Once again, we'll examine the code in pieces.

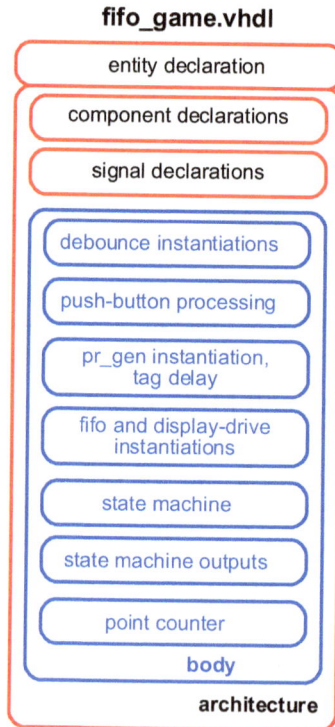

fifo_game.vhdl

entity declaration

component declarations

signal declarations

debounce instantiations

push-button processing

pr_gen instantiation, tag delay

fifo and display-drive instantiations

state machine

state machine outputs

point counter

body

architecture

First, the entity declaration.

```
entity fifo_game is
  port
  (
    clk            : in  std_logic;   ①
    pb_1           : in  std_logic;
    pb_2           : in  std_logic;   ②
    speed_select   : in  std_logic;   ③
    --                                 ④
    seven_seg_1    : out std_logic_vector(7 downto 0);
    seven_seg_2    : out std_logic_vector(7 downto 0);
    seven_seg_3    : out std_logic_vector(7 downto 0);
    seven_seg_4    : out std_logic_vector(7 downto 0);
    en1            : out std_logic;    ⑤
    en2            : out std_logic;    ⑥
    en3            : out std_logic;
    en4            : out std_logic;
    -- unused LEDs
    led            : out std_logic_vector(9 downto 0)
  );
end entity;
```

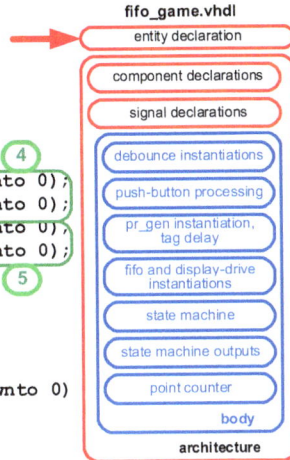

fifo_game.vhdl

- entity declaration
- component declarations
- signal declarations
- debounce instantiations
- push-button processing
- pr_gen instantiation, tag delay
- fifo and display-drive instantiations
- state machine
- state machine outputs
- point counter

body

architecture

1) this is the "next" push-button – reads another word from the FIFO;

2) this is the "affirm" push-button – indicates a tag guess;

3) the left-most slide switch – selects ~1 second or ~2 second step cycle periods;

4) the two single-segment displays – the FIFO input, and FIFO output;

5) the two character displays – FIFO fill level, and point count;

6) these are only used on the Digilent board.

Next are the first three of four total component declarations.

157

```
-- component declarations
component debounce is
port
  (
  clk           : in    std_logic;
  pb_in         : in    std_logic;          (1)
  --
  pb_debounced  : out   std_logic
  );
  end component;
```

```
component pr_gen is
port                                              (2)
  (
  clk          : in  std_logic;
  start        : in  std_logic;
  speed_select : in  std_logic;
  prbs         : out std_logic_vector(2 downto 0);
  tag          : out std_logic; --active for one full cycle time
  step         : out std_logic
  );
  end component;
```

```
component fifo is
port                                              (3)
  (
  clk       : in  std_logic;
  start     : in  std_logic;
  prbs_in   : in  std_logic_vector(2 downto 0);
  tag_in    : in  std_logic;
  wr        : in  std_logic;
  --
  rd        : in  std_logic;
  level     : out std_logic_vector(3 downto 0);
  prbs_out  : out std_logic_vector(2 downto 0);
  tag_out   : out std_logic
  );
  end component;
```

fifo_game.vhdl

- entity declaration
- component declarations
- signal declarations
- debounce instantiations
- push-button processing
- pr_gen instantiation, tag delay
- fifo and display-drive instantiations
- state machine
- state machine outputs
- point counter

body

architecture

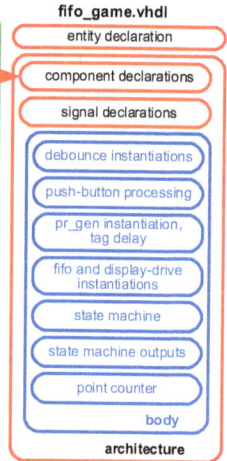

1) we've used this "debounce" component in previous projects;

2) the "pr_gen" component creates the various pseudo-random signals (the random segment, and the randomly generated tag signal), and the timing to establish a step cycle period;

3) the core FIFO. In addition to the actual pass-thru data (prbs and tag), the FIFO level is also provided.

And the last of the four components.

```
component display_drive is
port
  (
  clk           : in    std_logic;
  fifo_input    : in    std_logic_vector(2 downto 0);
  tag           : in    std_logic;
  end_game      : in    std_logic;
  fifo_output   : in    std_logic_vector(2 downto 0);
  fifo_level    : in    std_logic_vector(3 downto 0);
  point_count   : in    std_logic_vector(3 downto 0);
  --
  seven_seg_1   : out   std_logic_vector(7 downto 0);
  seven_seg_2   : out   std_logic_vector(7 downto 0);
  seven_seg_3   : out   std_logic_vector(7 downto 0);
  seven_seg_4   : out   std_logic_vector(7 downto 0);
  en1           : out   std_logic;
  en2           : out   std_logic;
  en3           : out   std_logic;
  en4           : out   std_logic
  );
end component;
```

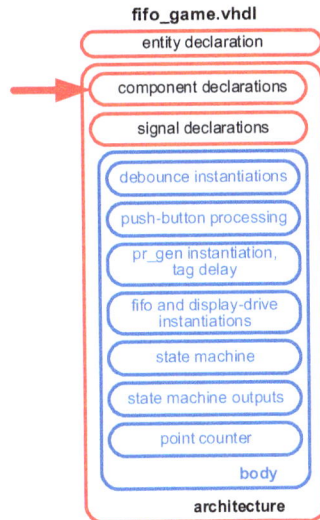

fifo_game.vhdl — entity declaration, component declarations, signal declarations, debounce instantiations, push-button processing, pr_gen instantiation, tag delay, fifo and display-drive instantiations, state machine, state machine outputs, point counter, body, architecture

Next comes the signal declarations.

```
type fifo_state_type is (                    ①
                          idle,
                          incr,
                          decr,
                          pop,
                          full,
                          clear
                        );
signal fifo_state : fifo_state_type;

signal pb_debncd_1       : std_logic;
signal pb_debncd_1_d1    : std_logic;
signal pb_debncd_2       : std_logic;
signal pb_debncd_2_d1    : std_logic;
signal pb_1_pls          : std_logic;
signal pb_2_pls          : std_logic;
signal pb_debncd_1_pol   : std_logic;
signal pb_debncd_2_pol   : std_logic;
signal start             : std_logic;
signal next_samp         : std_logic;
signal prbs              : std_logic_vector(2 downto 0);
signal affirm            : std_logic;
signal next_pls          : std_logic;
signal affirm_pls        : std_logic;
signal step              : std_logic;
signal tag           ②  : std_logic;
signal tag_d             : std_logic;
signal tag_out           : std_logic;
signal rd_fifo           : std_logic;
signal level             : std_logic_vector(3 downto 0);
signal prbs_out          : std_logic_vector(2 downto 0);
signal point_count       : unsigned(3 downto 0);
signal point_count_std   : std_logic_vector(3 downto 0);
signal point_cnt_clr     : std_logic;
signal point_cnt_incr    : std_logic;
signal point_cnt_decr    : std_logic;
signal end_game          : std_logic;
signal affirm_hold       : std_logic;
```

fifo_game.vhdl
- entity declaration
- component declarations
- signal declarations
- debounce instantiations
- push-button processing
- pr_gen instantiation, tag delay
- fifo and display-drive instantiations
- state machine
- state machine outputs
- point counter
- body
- architecture

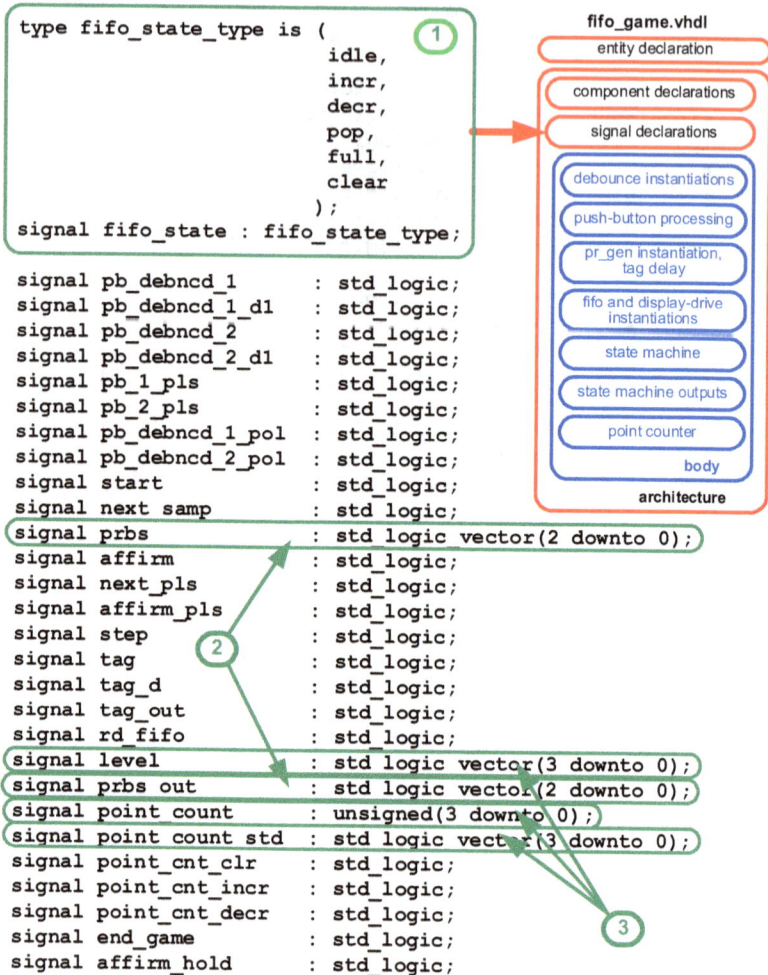

1) the state machine labels in the type form we're now familiar with;

2) the pseudo-random signals are 3 bits, which allow seven values for the seven segments of the displays;

3) whereas the FIFO level and points counter signals are four bits, since both span up to a value of 15.

We now move on to the architecture body, and the two debounce modules.

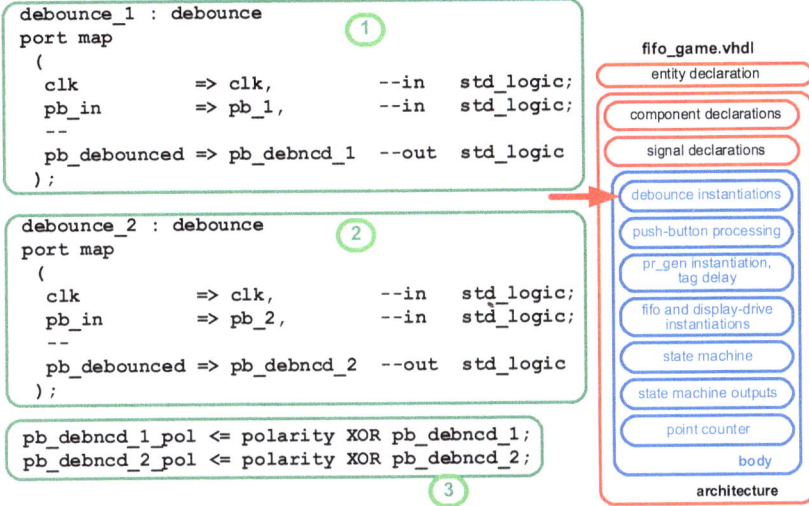

```
debounce_1 : debounce        ①
port map
(
  clk          => clk,         --in   std_logic;
  pb_in        => pb_1,        --in   std_logic;
  --
  pb_debounced => pb_debncd_1  --out  std_logic
);

debounce_2 : debounce        ②
port map
(
  clk          => clk,         --in   std_logic;
  pb_in        => pb_2,        --in   std_logic;
  --
  pb_debounced => pb_debncd_2  --out  std_logic
);

pb_debncd_1_pol <= polarity XOR pb_debncd_1;
pb_debncd_2_pol <= polarity XOR pb_debncd_2;
                                          ③
```

fifo_game.vhdl
- entity declaration
- component declarations
- signal declarations
- debounce instantiations
- push-button processing
- pr_gen instantiation, tag delay
- fifo and display-drive instantiations
- state machine
- state machine outputs
- point counter
- body
- architecture

1) the first debounce is for the "next" button,

2) and the second one is for the "affirm" button;

3) the push-buttons have different "senses" between the Terasic (low-active) and Digilent (high-active) boards, and we use the "polarity" signal, located in separate package files dedicated to each board, to normalize the sense as high-active.

Push-button processing next.

```
pb_proc : process(clk)
begin
  if rising_edge(clk) then
    pb_debncd_1_d1 <= pb_debncd_1_pol;
    pb_1_pls        <= pb_debncd_1_pol and NOT pb_debncd_1_d1;
    --
    pb_debncd_2_d1 <= pb_debncd_2_pol;
    pb_2_pls        <= pb_debncd_2_pol and NOT pb_debncd_2_d1;
    --
    if (pb_2_pls = '1') then
      affirm_hold <= '1';
    elsif (step = '1') then
      affirm_hold <= '0';
    end if;
    --
  end if;
end process;
```

fifo_game.vhdl

- entity declaration
- component declarations
- signal declarations
- debounce instantiations
- push-button processing
- pr_gen instantiation, tag delay
- fifo and display-drive instantiations
- state machine
- state machine outputs
- point counter

body

architecture

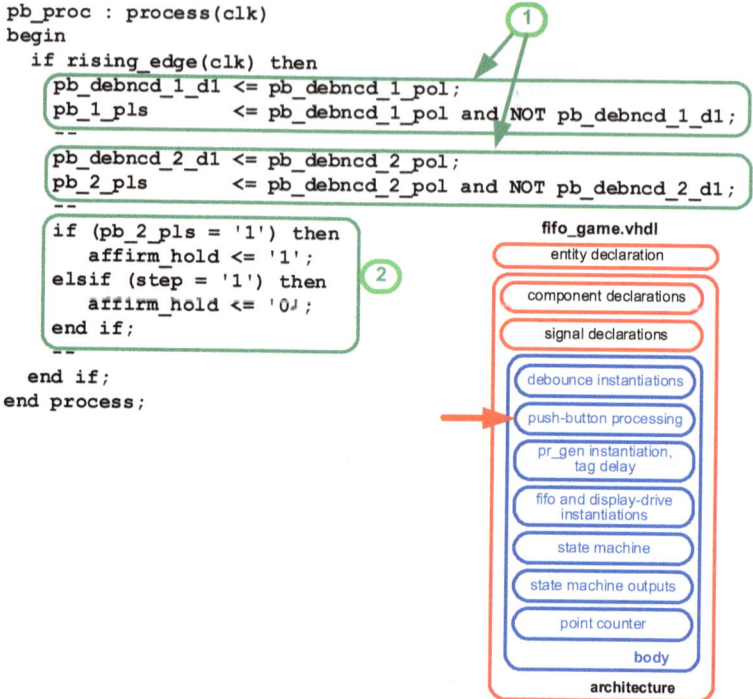

1) creating rising-edge detection pulses, as we've seen before;

2) here we have a set/reset latch. We set the "affirm_hold" signal active when the "affirm" button is pushed, and it stays active until the end of this step cycle period. As we'll see, this signal prevents multiple "affirm" activations within one step cycle.

Continuing with the pr_gen:

```
pr_gen_i : pr_gen
port map
  (
  clk          => clk,              --in  std_logic;
  start        => start,            --in  std_logic;
  speed_select => (NOT speed_select), --in  std_logic;
  prbs         => prbs,             --out std_logic_vector(2 downto 0);
  tag          => tag,              --out std_logic;
  step         => step              --out std_logic
  );

tag_delay : process(clk)
begin
  if rising_edge(clk) then
    if (step = '1') then
      tag_d <= tag;
    end if;
  end if;
end process;
```

fifo_game.vhdl
- entity declaration
- component declarations
- signal declarations
- debounce instantiations
- push-button processing
- pr_gen instantiation, tag delay
- fifo and display-drive instantiations
- state machine
- state machine outputs
- point counter
- body
- architecture

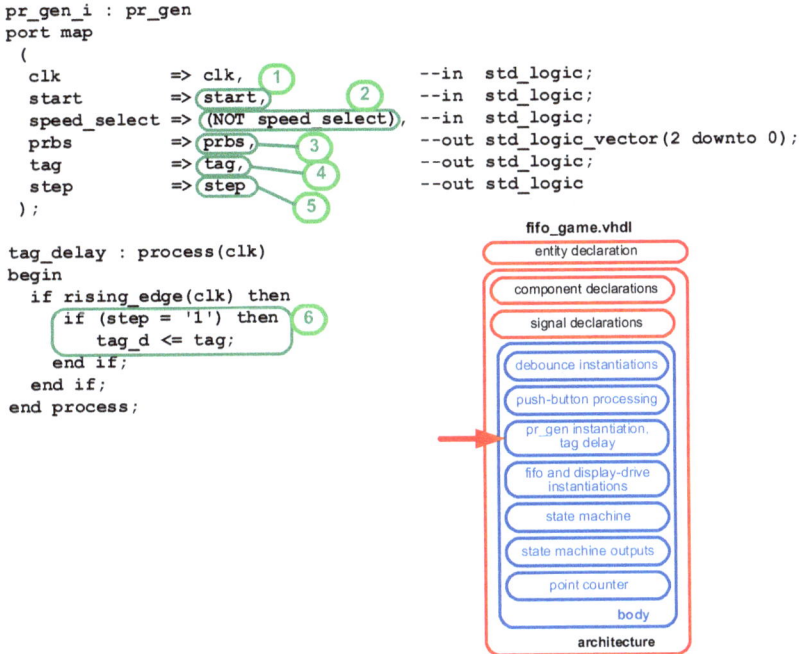

1) as we'll soon see, the "start" signal is created when the state machine passes through the "clear" state, and initiates just what it says – it starts a new game;

2) "speed_select" is sourced from the board's left-most slide-switch, and we invert it so that the "off" position selects the slower speed. Note here that VHDL allows a logic expression in the connection assignment;

3) the "prbs" signal comprises the random display segment that's both immediately displayed and also written into the FIFO for later game tagging display;

4) "tag" is the indication of a randomly generated game tag;

5) "step" defines the cycle period of the game, ~2 seconds when in slow mode, and ~1 second when in fast mode;

6) at the beginning of each step cycle we create a delayed version of "tag", which we'll see prevents a write to the FIFO during the flashing display tag notice.

The "fifo" and "display_drive" instantiations:

163

```
fifo_i : fifo
port map
  (
  clk       => clk,                      --in  std_logic;
  start     => start,                    --in  std_logic;
  prbs_in   => prbs,                     --in  std_logic_vector(2 downto 0);
  tag_in    => tag,                      --in  std_logic;
  wr        => (step AND NOT tag_d),     --in  std_logic;
  --
  rd        => (rd_fifo OR affirm_pls),  --in  std_logic;
  level     => level,                    --out std_logic_vector(3 downto 0);
  prbs_out  => prbs_out,                 --out std_logic_vector(2 downto 0);
  tag_out   => tag_out                   --out std_logic
  );

display_drive_i : display_drive
port map
  (
  clk           => clk,                  --in  std_logic;
  fifo_input    => prbs,                 --in  std_logic_vector(2 downto 0);
  tag           => tag_d,                --in  std_logic; -- flash display
  end_game      => end_game,             --in  std_logic;
  fifo_output   => prbs_out,             --in  std_logic_vector(2 downto 0);
  fifo_level    => level,                --in  std_logic_vector(3 downto 0);
  point_count   => point_count_std,      --in  std_logic_vector(3 downto 0);
  --
  seven_seg_1   => seven_seg_1,          --out std_logic_vector(7 downto 0);
  seven_seg_2   => seven_seg_2,          --out std_logic_vector(7 downto 0);
  seven_seg_3   => seven_seg_3,          --out std_logic_vector(7 downto 0);
  seven_seg_4   => seven_seg_4,          --out std_logic_vector(7 downto 0);
  en1           => en1,                  --out std_logic;
  en2           => en2,                  --out std_logic;
  en3           => en3,                  --out std_logic;
  en4           => en4                   --out std_logic
  );
```

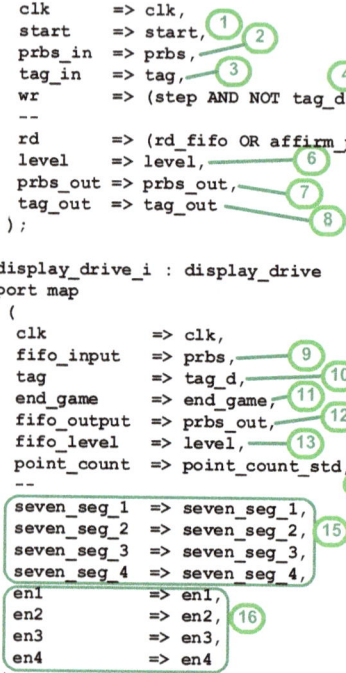

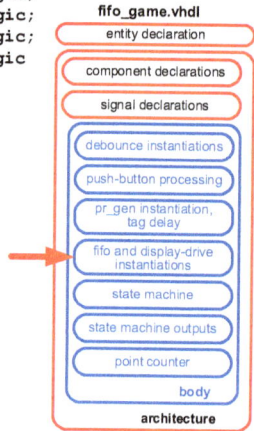

fifo_game.vhdl
entity declaration
component declarations
signal declarations
debounce instantiations
push-button processing
pr_gen instantiation, tag delay
fifo and display-drive instantiations
state machine
state machine outputs
point counter
body
architecture

1) "start" clears the FIFO at the beginning of a new game;

2) the random segment that's currently being presented to the first display;

3) the "tag" flag carried through the FIFO;

4) after the random tag segment value is displayed and written to the FIFO (during the step cycle when "tag" is active),

we then flash the two left-most displays for one step cycle, and the delayed "tag_d" prevents a write to the FIFO at the end of that period;

5) "rd_fifo" from "pop" state of the state machine, and we also read the next value from the FIFO when the user pushes the affirm button – this prevents the user from gaining additional points by simply pushing the affirm button multiple times;

6) the FIFO level to the state machine in the event of game-end (0xF), and for display to the user via "display_drive";

7) to "display_drive";

8) for comparison against a user's affirm choice;

9) for display on the second 7-segment unit;

10) used by "display_drive" to flash the first 7-segment display, indicating that the previous segment was a tag value;

11) used by "display_drive" when a game ends to flash the first first two 7-segment displays;

12) the segment to be displayed on the first 7-segment display;

13) the FIFO level displayed on the third 7-segment display;

14) the user's point score;

15) drives the four 7-segment displays;

16) used by the Digilent board as we've seen before.

The following code for the state machine directly implements the diagram we saw previously. Note that we are not able to use the signal name "next", since that's a VHDL keyword – thus, "next_samp" (for "next sample").

```
state_machine : process(clk)
begin
  if rising_edge(clk) then
    case (fifo_state) is
      when idle  => if (affirm_pls = '1') then
                      if (tag_out = '1') then       fifo_state <= incr;
                      else                          fifo_state <= decr;
                      end if;
                    elsif (next_pls = '1') then    fifo_state <= pop;
                    elsif (step = '1') then
                      if (level = X"F") then        fifo_state <= full;
                      elsif ( [next_samp = '1']
                             AND affirm = '1'
                           ) then                   fifo_state <= clear;
                      else                          fifo_state <= idle;
                      end if;
                    end if;
      when incr  =>                                 fifo_state <= pop;
      when decr  =>                                 fifo_state <= pop;
      when pop   =>                                 fifo_state <= idle;
      when full  =>  if ( [next_samp = '1']
                         AND affirm = '1'
                       ) then                       fifo_state <= clear;
                     end if;
      when clear =>                                 fifo_state <= idle;
      --
      when others => fifo_state <= idle;
    end case;
  end if;
end process;
```

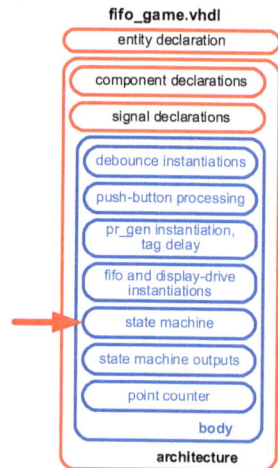

fifo_game.vhdl

- entity declaration
- component declarations
- signal declarations
  - debounce instantiations
  - push-button processing
  - pr_gen instantiation, tag delay
  - fifo and display-drive instantiations
  - → state machine
  - state machine outputs
  - point counter
  - **body**

architecture

For clarity, since we have many actions driven by the state machine, we use a separate case statement for the state machine outputs.

```
state_machine_outputs : process(clk)
begin
   if rising_edge(clk) then
      start          <= '0';      A
      point_cnt_clr  <= '0';
      point_cnt_incr <= '0';
      point_cnt_decr <= '0';
      rd_fifo        <= '0';
      end_game       <= '0';
      --
      case (fifo_state) is
         when idle  => null;
         when incr  => point_cnt_incr <= '1';   B
         when decr  => point_cnt_decr <= '1';
         when pop   => rd_fifo        <= '1';
         when full  => point_cnt_clr  <= '1';
                       end_game       <= '1';
         when clear => start          <= '1';
                       point_cnt_clr  <= '1';
         --
         when others => null;
      end case;
   end if;
end process;
```

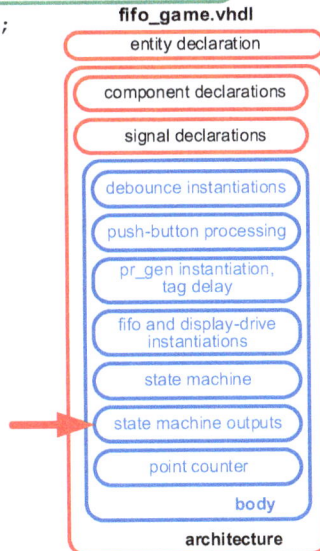

fifo_game.vhdl
- entity declaration
- component declarations
- signal declarations
  - debounce instantiations
  - push-button processing
  - pr_gen instantiation, tag delay
  - fifo and display-drive instantiations
  - state machine
  - state machine outputs
  - point counter
- body
- architecture

Although the entirety of this process statement takes affect at precisely the rising edge of the clock, since the body of a process statement is logically interpreted from beginning to end, any assignments that are not made later (B), remain assigned at zero (A). The effect of this is that state machine outputs are active only during the associated state, which is a single clock pulse for all of them except the "full" state.

167

Blaine C. Readler

We can note here that all these assignments could have been made within the state machine case statement itself, but this tends to add clutter there, and by separating them, the flow of the state machine is much easier to see.

The last piece of the top file architecture is the point counter, the sort of counting that we've seen before, noting:

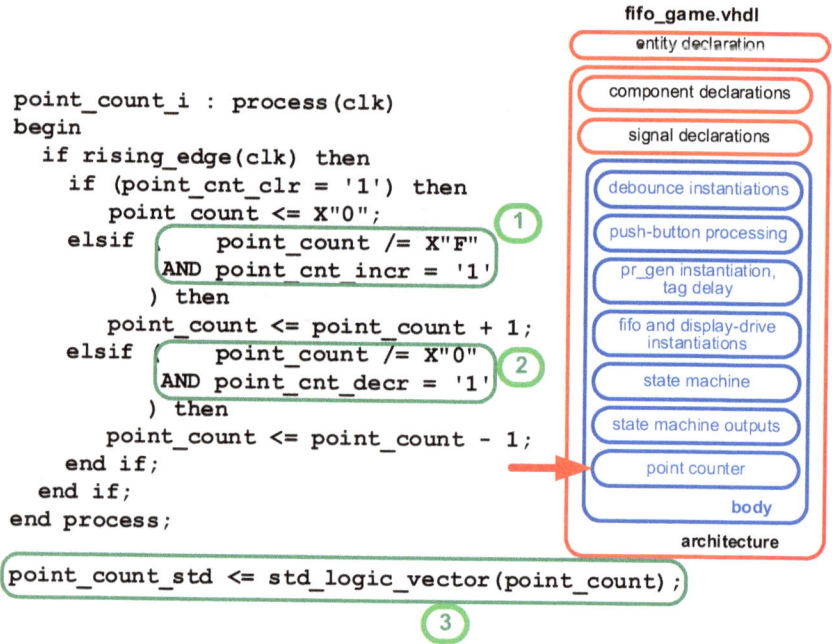

```
point_count_i : process(clk)
begin
  if rising_edge(clk) then
    if (point_cnt_clr = '1') then
      point_count <= X"0";
    elsif (    point_count /= X"F"      1
            AND point_cnt_incr = '1'
          ) then
      point_count <= point_count + 1;
    elsif (    point_count /= X"0"      2
            AND point_cnt_decr = '1'
          ) then
      point_count <= point_count - 1;
    end if;
  end if;
end process;
```

```
point_count_std <= std_logic_vector(point_count);   3
```

fifo_game.vhdl

- entity declaration
- component declarations
- signal declarations
- debounce instantiations
- push-button processing
- pr_gen instantiation, tag delay
- fifo and display-drive instantiations
- state machine
- state machine outputs
- point counter

body
architecture

1) the counter does not roll over, but freezes at "F" when incrementing;
2) similarly, the counter does not roll under, but freezes at "0" when decrementing;
3) we convert the count to std_logic (from unsigned) for delivery to the "display_drive" module.

That concludes the top file, and we now move on to look into the instantiated modules, or at least those that we haven't covered previously.

We begin with the "pr_gen" module.

pr_gen.vhdl

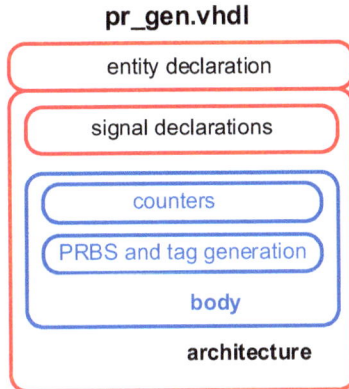

The entity, as we've explained in the top file instantiation:

```
entity pr_gen is
  port
    (
    clk          : in  std_logic;
    start        : in  std_logic;
    speed_select : in  std_logic;
    prbs         : out std_logic_vector(2 downto 0);
    tag          : out std_logic;
    step         : out std_logic
    );
end entity;
```

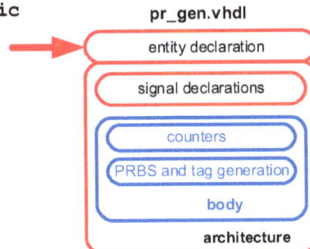

The signal declarations:

```
architecture Behavioral of pr_gen is

    signal prs        : std_logic_vector(11 downto 1);
    signal step_lcl   : std_logic;
    signal tag_lcl    : std_logic;                        (1)
    signal tag_lcl_d1 : std_logic;
    signal tag_lcl_d2 : std_logic;
    signal tag_lcl_d3 : std_logic;                    (2)
    signal step_cnt   : unsigned(27 downto 0)   (3)    (4)
    signal load_cnt   : unsigned(10 downto 0)
    signal prs_prev   : std_logic_vector(3 downto 1);
    signal tag_check  : std_logic_vector(2 downto 0);  (5)
    signal prs_alt    : std_logic_vector(2 downto 0);  (6)
```

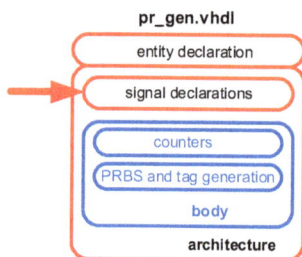

pr_gen.vhdl

entity declaration

→ signal declarations

counters

PRBS and tag generation

body

architecture

1) we've seen a pseudo-random generator in the previous project, and this one uses a longer sequence;

2) 28 bits allows our step cycle to be up to five seconds;

3) and the "load_cnt" is eleven bits to match the pseudo-random generator (as we'll see);

4) although not necessary, we maintain the "3 downto 1" instead of "2 downto 0" for clarity;

5) three bits, since our tag is a 3-bit value (1:8 chance of a tag);

6) at times we need an alternate tag value (as we'll see).

Our two counters:

```
begin

    step_cnt_proc : process(clk)
    begin
        if rising_edge(clk) then
            if (step_cnt = X"0000000") then
                if (speed_select = '0') then
                    step_cnt <= step_load_slow;
                else
                    step_cnt <= step_load_fast;
                end if;
                --
                step_lcl <= '1';
            else
                step_cnt <= step_cnt - 1;
                step_lcl <= '0';
            end if;
        end if;
    end process;

    load_cnt_proc : process(clk)
    begin
        if rising_edge(clk) then
            load_cnt <= load_cnt + 1;
        end if;
    end process;
```

pr_gen.vhdl

entity declaration

signal declarations

counters

PRBS and tag generation

body

architecture

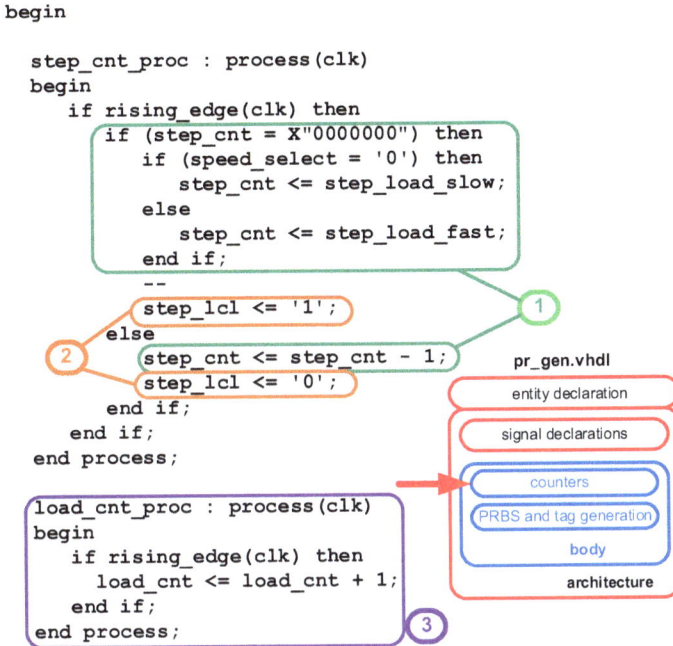

1) we've seen this sort of thing previously – when the down-counter reaches zero, we reload it. In this case, by using two different size starting values, we achieve two different step cycle periods, and thus two difficulties for the game. We use constants for reload values ("step_load_slow" and "step_load_fast") from two board-specific packages since the basic board clock is different between the Terasic and Digilent boards;

2) "step_lcl" defines the beginning of the next step cycle, and is a pulse one clock wide;

3) "load_cnt" is a free-running counter that, as we'll see next, provides a random initial load of the pseudo-random generator at the beginning of each new game.

On to the rest of "pr_gen":

```
prs_process : process(clk)
begin
    if rising_edge(clk) then
        if (start = '1') then --load seeds          1
            prs <= std_logic_vector(load_cnt);
        elsif (step_lcl = '1') then    -- x^11 + x^9 + x^1
            prs        <=    prs(1)
                          & prs(11)                    2
                          & (prs(10) xor prs(1))
                          & prs(9 downto 2);
            prs_prev <= prs(3 downto 1);    3
        end if;
        --
        if (step_lcl = '1') then
            if (    tag_check = "111"                  4
                AND tag_lcl    /= '1' --previous
                AND tag_lcl_d1 /= '1' --two prior
                AND tag_lcl_d2 /= '1' --three prior
                AND tag_lcl_d3 /= '1' --four prior
                ) then
                tag_lcl <= '1';
            else
                tag_lcl <= '0';
            end if;
            --                              pr_gen.vhdl
            tag_lcl_d1 <= tag_lcl;      6   entity declaration
            tag_lcl_d2 <= tag_lcl_d1;
            tag_lcl_d3 <= tag_lcl_d2;       signal declarations
        end if;
    end if;                                 counters
end process;
                                  5         PRBS and tag generation
tag_check <= (prs(11) & prs(8) & prs(5));   body
                                            architecture
prs_alt <= (prs(9) & prs(5) & prs(2));   7
prbs    <= prs(3 downto 1) when (prs(3 downto 1) /= prs_prev)
                     else
           prs(3 downto 1) XOR prs_alt;         8
--
step <= step_lcl;   9
tag  <= tag_lcl;
```

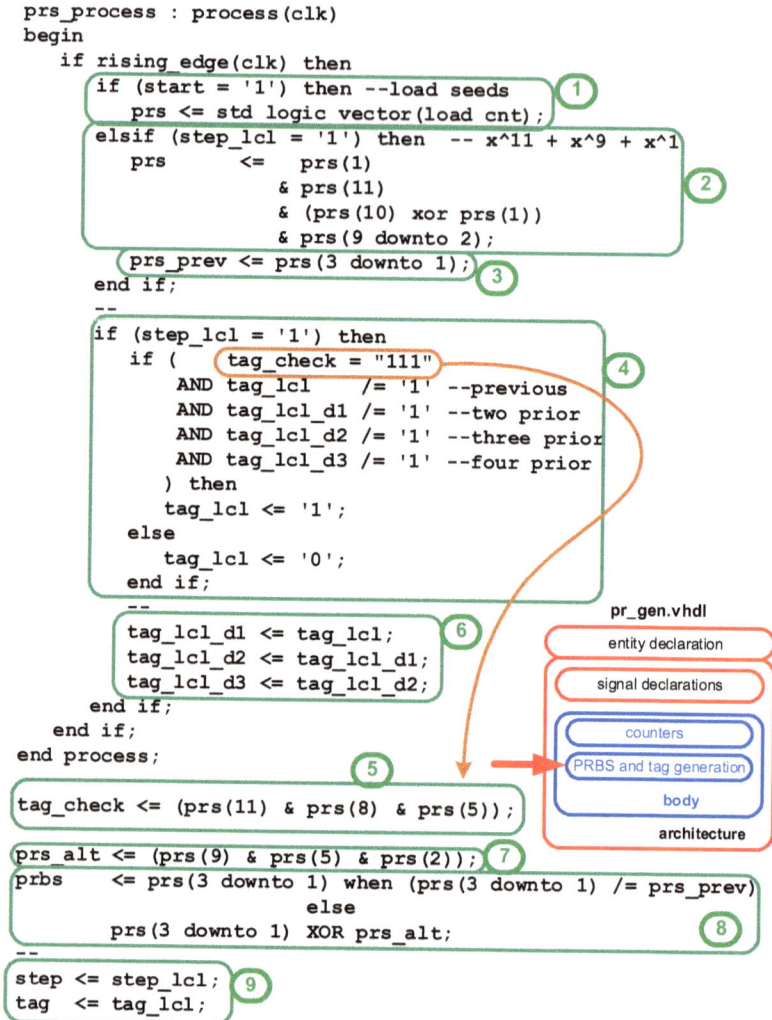

1) this is where the random "load_cnt" loads the pseudo-random generator at the beginning of each new game;

2) we use the same method in creating a pseudo-random value as we did previously in the match game, except that here, the sequence is 11 bits long instead of 9 bits, providing 2^{11}, or 2,047 unique numbers instead of the match game's 511. A new pseudo-random value is generated each step cycle;

3) each step cycle we save off the current pseudo-random value that, as we'll see, we use the next step cycle to avoid using

the same pseudo-random segment value twice in a row, which would result in the display segment not changing from one step cycle to the next;

4) similarly, we want to avoid issuing tags too often, and here we don't set "tag_lcl" if there was a previous tag issued within the last four step cycles;

5) "tag_check", which comprises a new potential tag, is derived from three separate bits of the pseudo-random generator;

6) and here's where we develop the delayed tags used above, noting the the assignments are made – essentially a shift register – at the beginning of each step cycle;

7) three other separate pseudo-random generator bits ("prs_alt") that we use . . .

8) if the current pending pseudo-random segment ("prs(3 downto 1")) is going to be the same as the last one ("prs_prev" that we saved off above). If they are the same, we use "prs_alt" to more or less scramble the pending pseudo-random segment, ensuring that it will be different;

9) VHDL doesn't allow us to use output signals from the component (as outputs in the entity list above), so we have to develop local signals (i.e., "*_lcl") to use in the body that we then reassign as the actual outputs – one of many seemingly silly VHDL inconveniences.

We move on to the FIFO – "fifo.vhdl", which, as we'll see, is a special type of memory.

fifo.vhdl

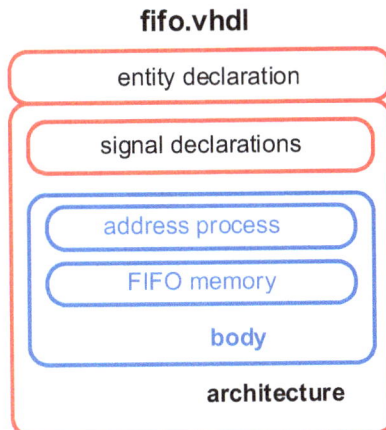

As always, we begin with the component entity declaration.

```
entity fifo is
  port
    (
    clk         : in  std_logic;
    start       : in  std_logic;        ①                    ②
    prbs_in     : in  std_logic_vector(2 downto 0);
    tag_in      : in  std_logic;
    wr          : in  std_logic;
    --
    rd          : in  std_logic;                              ③
    level       : out std_logic_vector(3 downto 0);
    prbs_out    : out std_logic_vector(2 downto 0);
    tag_out     : out std_logic                              ④
    );
  end entity;
```

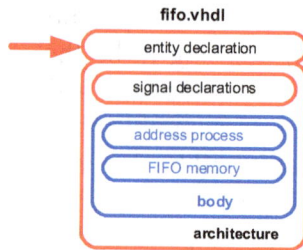

fifo.vhdl

entity declaration

signal declarations

address process

FIFO memory

body

architecture

1) as we'll soon see, we clear the FIFO at the beginning of each game;

2) the inputs to the FIFO, three bits of pseudo-random segment, and one tag bit, for a total of four bits;

3) just what it indicates, this is the level of the FIFO, i.e., how many words it contains;

4) the FIFO outputs – the same two inputs, but delayed multiple step cycles as they make their way through the FIFO.

Signal declarations:

```
architecture Behavioral of fifo is
```

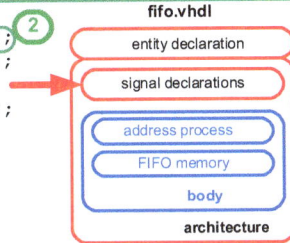

```
type array_16x4 is array(0 to 15) of std_logic_vector(3 downto 0);
signal mem : array_16x4;
--
signal level_cnt    : unsigned(3 downto 0);
signal wr_adr       : unsigned(3 downto 0);
signal wr_adr_int   : integer;
signal rd_adr       : unsigned(3 downto 0);
signal rd_adr_int   : integer;
signal tag_out_lcl  : std_logic;
```

1) since a FIFO is essentially a special type of memory, we implement it using the same array method as we did in the previous memory project – in this case, four bits wide, and sixteen deep;

2) since the level amount is tracked by a counter, we must use "unsigned" as the signal type. The write and read addresses are counts as well, as we've seen before.

The first part of the body, where we generate the write addresses (FIFO inputs), and read addresses (FIFO outputs), and also track the FIFO level.

175

```
address_proc : process(clk)
begin
   if rising_edge(clk) then
      if (start = '1') then          1
         wr_adr    <= X"4";
         rd_adr    <= X"0";
         level_cnt <= X"4";
      end if;
      --
      if (wr = '1') then             2
         wr_adr <= wr_adr + 1;
      end if;
      --
      if (    level_cnt > 4
          AND rd = '1'               3
          ) then
          rd_adr <= rd_adr + 1;
      end if;
                                     4
      if (    level_cnt > 4
          AND rd = '1'
          ) then
          level_cnt <= level_cnt - 1;
      elsif (wr = '1') then
          level_cnt <= level_cnt + 1;
      end if;
   end if;
end process;
```

fifo.vhdl

entity declaration

signal declarations

address process

FIFO memory

body

architecture

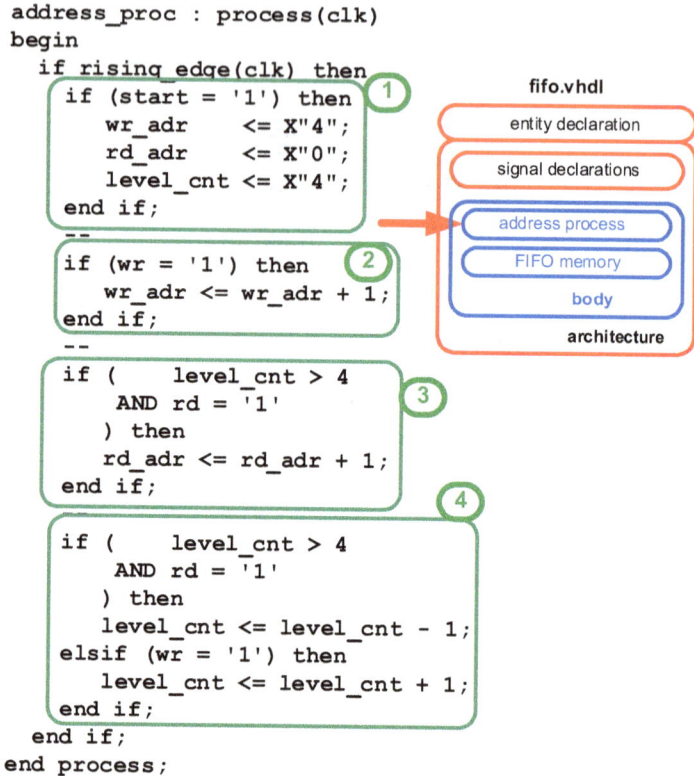

1) at the beginning of a game ("start") we clear the FIFO by setting the write address to zero, but we also fix the read address, and thus also the FIFO level, to four. The reason for this is to prevent the user from quickly reading the FIFO to the point where a tag output appears immediately one step after an observed input tag, making the game too easy;

2) each write to the FIFO, we increment the write address, and thus subsequent inputs are laid in one after the other, filling the FIFO. Note that the address count will roll over when it reaches 0xF. At this point, hopefully the user will have read some outputs, otherwise this would constitute a FIFO overflow;

3) incrementing the read address pulls values from the FIFO, following behind the FIFO entry write addresses. However, using the FIFO fill level, we don't let the user get less than four entries closer to the written inputs. Like the write

addresses, the read address rolls over, chasing the write address around and around.

4) the FIFO level counter simply increments and decrements with each FIFO write and read, respectively.

And finally, the memory itself.

```
wr_adr_int <= to_integer(unsigned(wr_adr));
rd_adr_int <= to_integer(unsigned(rd_adr));
                                              (1)
memory_proc : process(clk)
begin
    if rising_edge(clk) then               (2)
        if (wr = '1') then
            mem(wr_adr_int) <= (prbs_in & tag_in);
        end if;
                                              (3)
        --
        prbs_out      <= mem(rd_adr_int)(3 downto 1);
        tag_out_lcl   <= mem(rd_adr_int)(0);
    end if;
end process;
                                    (4)
level    <= std_logic_vector(level_cnt);
tag_out <= tag_out_lcl;

end architecture Behavioral;
```

fifo.vhdl

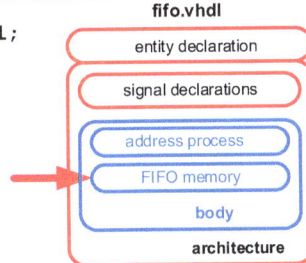

1) as with the original memory in the previous project, we convert the addresses to integers to use as memory array indexes;

2) we concatenate (join together) the three-bit prbs value with the one-bit tag, so that the FIFO is written with one four-bit value,

3) and so when we read the value from the FIFO memory, we must disassemble the four-bit value into separate prbs and tag signals;

177

4) as we've already seen, to output these signals from this component, we must convert the unsigned level count to std_logic, and use local signals for the tag.

The "display_drive" component, in addition to presenting translated numeric/hex characters to the last two displays (FIFO level and point count), also translates the three-bit pseudo-random game values to 8 bits of individual segment locations for the first two displays (FIFO in and FIFO out). As in other projects, differences between the Terasic and Digilent boards on how the 7-segment displays are driven require two different "display_drive" components, each in their own board IP area.

Terasic "display_drive"

display_drive.vhdl

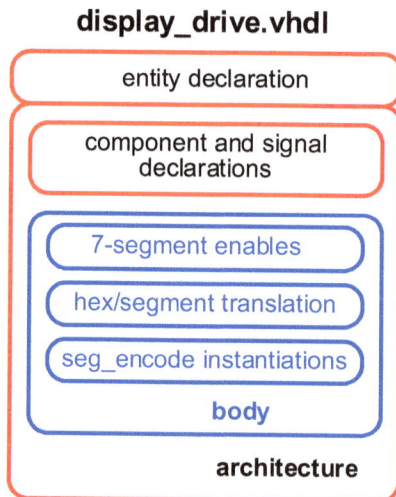

First, the entity:

```
entity display_drive is
  port
  (
    clk            : in    std_logic;               ①
    fifo_input     : in    std_logic_vector(2 downto 0);
    tag            : in    std_logic; ②
    end_game       : in    std_logic;    ③                  ④
    fifo_output    : in    std_logic_vector(2 downto 0);
    fifo_level     : in    std_logic_vector(3 downto 0); ⑤
    point_count    : in    std_logic_vector(3 downto 0); ⑥
    --
    seven_seg_1    : out   std_logic_vector(7 downto 0),
    seven_seg_2    : out   std_logic_vector(7 downto 0); ⑦
    seven_seg_3    : out   std_logic_vector(7 downto 0);
    seven_seg_4    : out   std_logic_vector(7 downto 0);
    en1            : out   std_logic;
    en2            : out   std_logic; ⑧
    en3            : out   std_logic;
    en4            : out   std_logic
  );
end entity;
```

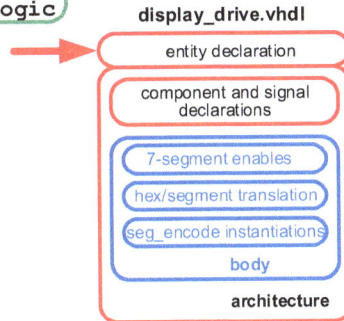

display_drive.vhdl

entity declaration

component and signal declarations

7-segment enables

hex/segment translation

seg_encode instantiations

body

architecture

1) the left-most 7-segment display;

2) the "tag" indication causes the left-most display to flash the tag signal to the user;

3) causes the two left-most displays to continually flash, indicating game-over;

4) the second 7-segment display from the left;

5) the third 7-segment display from the left;

6) the last 7-segment display, fourth from the left;

7) the 7-segment display outputs;

8) the 7-segment displays are continually enabled for the Terasic board.

The component and signal declarations. The same "seg_encode" that we've used previously:

179

```
architecture Behavioral of display_drive is

  component seg_encode
  port
   (
    hex_in      : in   std_logic_vector(3 downto 0);
    seven_seg   : out  std_logic_vector(7 downto 0)
   );
  end component;

  signal blink_cnt     : unsigned(27 downto 0);
  signal encode3_input : std_logic_vector(3 downto 0);

begin
```

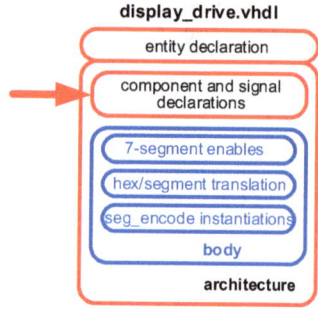

display_drive.vhdl

On the Terasic boards, the 7-segment displays are continually enabled.

```
begin

  en1 <= '0'; -- fixed active for Terasic
  en2 <= '0'; -- "
  en3 <= '0'; -- "
  en4 <= '0'; -- "
```

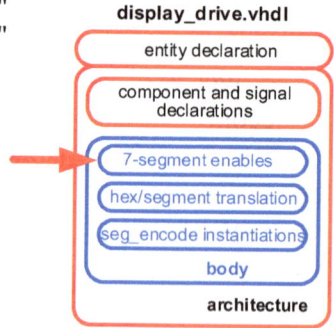

display_drive.vhdl

Here we translate the hex 3-bit PRS values to the individual eight segments of the first two 7-segment displays.

```
timer : process(clk)
begin
   if rising_edge(clk) then
      if (blink_cnt = X"0000000") then
         blink_cnt <= X"1000000";            ①
      else
         blink_cnt <= blink_cnt - 1;
      end if;
      --                                      ②
      if (end_game = '1') then
         seven_seg_1 <= (others => blink_cnt(23));
         seven_seg_2 <= (others => blink_cnt(23));
      elsif (tag = '1') then                  ③
         seven_seg_1 <= (others => blink_cnt(23));
      else
         case(fifo_input) is
            when "000" => seven_seg_1 <= "11111110";
            when "001" => seven_seg_1 <= "11111101";
            when "010" => seven_seg_1 <= "11111011";  ④
            when "011" => seven_seg_1 <= "11110111";
            when "100" => seven_seg_1 <= "11101111";
            when "101" => seven_seg_1 <= "11011111";
            when "110" => seven_seg_1 <= "10111111";
            when "111" => seven_seg_1 <= "01111111";
            when others => null;
         end case;
         --
         case(fifo_output) is
            when "000" => seven_seg_2 <= "11111110";
            when "001" => seven_seg_2 <= "11111101";
            when "010" => seven_seg_2 <= "11111011";
            when "011" => seven_seg_2 <= "11110111";
            when "100" => seven_seg_2 <= "11101111";
            when "101" => seven_seg_2 <= "11011111";
            when "110" => seven_seg_2 <= "10111111";
            when "111" => seven_seg_2 <= "01111111";
            when others => null;
         end case;
      end if;
   end if;
end process;
```

display_drive.vhdl

entity declaration

component and signal declarations

7-segment enables

hex/segment translation

seg_encode instantiations

body

architecture

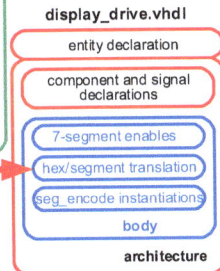

1) a free-running counter that toggles the first two 7-segment displays on and off as a blink to indicate either a tag (just the first display), or end-game (the first two displays);

2) we blink both of the first two displays to indicate that the game has ended. The MS bit of the blink count toggles at a rate of 2^{23}, or about half a second. We find a new VHDL feature here: "(others => *)" represents a std_logic_vector that has all the bits set to "*". This form allows us to assign a signal that's

either all zeros (when "blink_cnt(23) is low), or all ones (when "blink_cnt(23) is high). This is handy when either we have a long string of bits to assign, or when the field width is not easy to determine;

3) similarly, we blink the first display (the FIFO input) to indicate to the user that the previous segment was a tag;

4) lastly, we have the translations from the three-bit PRS values to individual segments of the displays. You can see that each combination of PRS results in one unique segment being lit, remembering that the segments are low-active.

Finally we come to the "seg_encode" components that we've seen before, translating the hex count values to 7-segment characters"

```
encode3_input <= X"0" when end_game = '1' else fifo_level;

   seg_encode_3 : seg_encode
   port map
     (
     hex_in      => encode3_input,
     seven_seg   => seven_seg_3
     );

   seg_encode_4 : seg_encode
   port map
     (
     hex_in      => point_count,
     seven_seg   => seven_seg_4
     );

end architecture Behavioral;
```

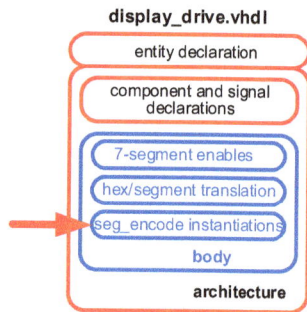

display_drive.vhdl
entity declaration
component and signal declarations
7-segment enables
hex/segment translation
seg_encode instantiations
body
architecture

The third display shows the FIFO level, except a zero when the game ends, while the first two displays are flashing.

Here's how the game is laid out on the Terasic board.

1) the "next" push-button;

2) the "affirm" push-button;

3) the game difficulty selection – off (towards the board edge) cycles each game step approximately once every two seconds; on cycles each game step approximately once a second.

A) the FIFO input segment – the segment prior to the flashing "tag" indication that must be remembered;

B) the FIFO output segment – when this matches the tagged one, pushing the "affirm" push-button scores one point. Pushing the "affirm" button otherwise, loses a point;

C) the FIFO level – never falls below four. Reaching "F" ends the game;

D) the point score.

Blaine C. Readler

Digilent "display_drive"

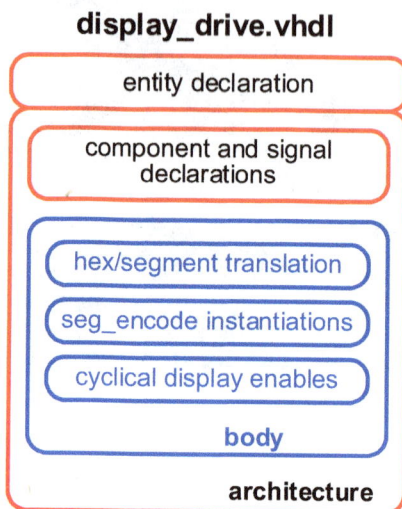

display_drive.vhdl

> entity declaration

> component and signal declarations

> > hex/segment translation
> >
> > seg_encode instantiations
> >
> > cyclical display enables
> >
> > **body**
>
> **architecture**

First, the entity:

```
entity display_drive is
  port
    (
    clk            : in    std_logic;                      ①
    fifo_input     : in    std_logic_vector(2 downto 0);
    tag            : in    std_logic;     ②
    end_game       : in    std_logic;        ③              ④
    fifo_output    : in    std_logic_vector(2 downto 0);
    fifo_level     : in    std_logic_vector(3 downto 0);   ⑤
    point_count    : in    std_logic_vector(3 downto 0);   ⑥
    --
    seven_seg_1    : out   std_logic_vector(7 downto 0);
    seven_seg_2    : out   std_logic_vector(7 downto 0);   ⑦
    seven_seg_3    : out   std_logic_vector(7 downto 0);
    seven_seg_4    : out   std_logic_vector(7 downto 0);
    en1            : out   std_logic;
    en2            : out   std_logic;        ⑧
    en3            : out   std_logic;
    en4            : out   std_logic
    );                                        display_drive.vhdl
end entity;
```

display_drive.vhdl

entity declaration

component and signal declarations

hex/segment translation

seg_encode instantiations

cyclical display enables

body

architecture

1) the left-most 7-segment display;

2) the "tag" indication causes the left-most display to flash the tag signal to the user;

3) causes the two left-most displays to continually flash, indicating game-over;

4) the second 7-segment display from the left;

5) the third 7-segment display from the left;

6) the last 7-segment display, fourth from the left;

7) the 7-segment display outputs;

8) the 7-segment displays require continuous cyclical enabling for the Digilent board.

The component and signal declarations. The same "seg_encode" that we've used previously:

```
architecture Behavioral of display_drive is

  component seg_encode
  port
   (
    hex_in      : in    std_logic_vector(3 downto 0);
    seven_seg   : out   std_logic_vector(7 downto 0)
   );
  end component;

  signal blink_cnt     : unsigned(27 downto 0);
  signal encode3_input : std_logic_vector(3 downto 0);

begin
```

display_drive.vhdl

entity declaration

→ component and signal declarations

hex/segment translation

seg_encode instantiations

cyclical display enables

body

architecture

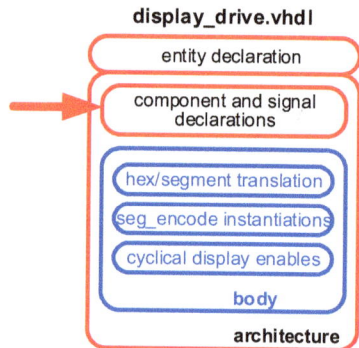

Here we translate the hex 3-bit PRS values to the individual eight segments of the first two 7-segment displays:

```
timer : process(clk)
begin
   if rising_edge(clk) then
      if (blink_cnt = X"0000000") then
         blink_cnt <= X"1000000";          (1)
      else
         blink_cnt <= blink_cnt - 1;
      end if;
      --                                    (2)
      if (end_game = '1') then
         seven_seg_1 <= (others => blink_cnt(23));
         seven_seg_2 <= (others => blink_cnt(23));
      elsif (tag = '1') then                (3)
         seven_seg_1 <= (others => blink_cnt(23));
      else
         case(fifo_input) is
           when "000" => seven_seg_1 <= "11111110";
           when "001" => seven_seg_1 <= "11111101";
           when "010" => seven_seg_1 <= "11111011";  (4)
           when "011" => seven_seg_1 <= "11110111";
           when "100" => seven_seg_1 <= "11101111";
           when "101" => seven_seg_1 <= "11011111";
           when "110" => seven_seg_1 <= "10111111";
           when "111" => seven_seg_1 <= "01111111";
           when others => null;
         end case;
         --
         case(fifo_output) is
           when "000" => seven_seg_2 <= "11111110";
           when "001" => seven_seg_2 <= "11111101";
           when "010" => seven_seg_2 <= "11111011";
           when "011" => seven_seg_2 <= "11110111";
           when "100" => seven_seg_2 <= "11101111";
           when "101" => seven_seg_2 <= "11011111";
           when "110" => seven_seg_2 <= "10111111";
           when "111" => seven_seg_2 <= "01111111";
           when others => null;
         end case;
      end if;
   end if;
end process;
```

display_drive.vhdl
entity declaration
component and signal declarations
hex/segment translation
seg_encode instantiations
cyclical display enables
body
architecture

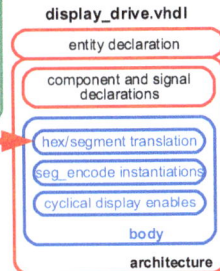

1) a free-running counter that toggles the first two 7-segment displays on and off as a blink to indicate either a tag (just the first display), or end-game (the first two displays);

2) we blink both of the first two displays to indicate that the game has ended. The MS bit of the blink count toggles at a rate of 2^{23}, or about half a second. We find a new VHDL feature here: "(others => *)" represents a std_logic_vector that has all the bits set to "*". This form allows us to assign a signal that's

Blaine C. Readler

either all zeros (when "blink_cnt(23) is low), or all ones (when "blink_cnt(23) is high). This is handy when either we have a long string of bits to assign, or when the field width is not easy to determine;

3) similarly, we blink the first display (the FIFO input) to indicate to the user that the previous segment was a tag;

4) lastly, we have the translations from the three-bit PRS values to individual segments of the displays. You can see that each combination of PRS results in one unique segment being lit, remembering that the segments are low-active.

Next we come to the "seg_encode" components that we've seen before, translating the hex count values to 7-segment characters:

```
encode3_input <= X"0" when end_game = '1' else fifo_level;

seg_encode_3 : seg_encode
port map
  (
  hex_in      => encode3_input,
  seven_seg   => seven_seg_3
  );

seg_encode_4 : seg_encode
port map
  (
  hex_in      => point_count,
  seven_seg   => seven_seg_4
  );
```

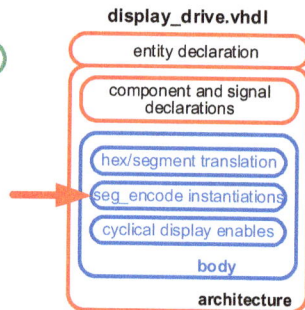

display_drive.vhdl

entity declaration

component and signal declarations

hex/segment translation
seg_encode instantiations
cyclical display enables
body
architecture

The third display shows the FIFO level, except a zero when the game ends, while the first two displays are flashing.

As we saw previously, the four 7-segment displays on the Digilent board share one 8-bit display drive bus (here, seven_seg_1), and each display is selected by its own enable (en1, en2, en3, en4). This section of code cycles around the four displays, enabling each in turn.

188

```
-- Cycle through the four segments.
main_proc : process(clk)
begin
   if rising_edge(clk) then    ①
      ms_cnt <= ms_cnt + 1;   -- ~~1.3ms at 100MHz
      --
      if (ms_cnt = '0' & X"0000") then    ②
         seg_sel <= seg_sel + 1;
      end if;

   end if;
end process;                    ③

-- 7-segment enables are low-active
en1 <= '0' when seg_sel = "00" else '1';
en2 <= '0' when seg_sel = "01" else '1';
en3 <= '0' when seg_sel = "10" else '1';
en4 <= '0' when seg_sel = "11" else '1';    ④

seven_seg_1 <= seg_1 when seg_sel = "00" else
               seg_2 when seg_sel = "01" else
               seg_3 when seg_sel = "10" else
               seg_4;

seven_seg_2 <= X"00";    ⑤
seven_seg_3 <= X"00";
seven_seg_4 <= X"00";

end architecture Behavioral;
```

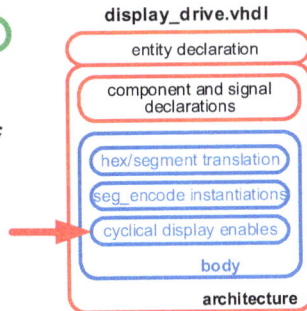

display_drive.vhdl

entity declaration

component and signal declarations

hex/segment translation

seg_encode instantiations

cyclical display enables

body

architecture

1) the counter, "ms_cnt", continually increments, cycling from 0x00 through 0x1FFF (0 to 131,071), and then starting over. At the 100MHz clock rate, the complete cycle time is about 1.3 milliseconds;

2) each time the 1.3ms cycle starts over (passes through zero) the 2-bit "seg_sel" signal increments, counting from "00" to "11", also starting over when it reaches "11";

3) as seg_sel cycles through its four counts, it activates the four displays in turn, noting that the display enables are low-active;

189

4) at the same time, the translated hex values from the memories (seg_1, etc.) are mux'd (selected) to the shared bus, "seven_seg_1". Thus, the contents of the two memories are presented to the 7-segment displays in turn, and each display is concurrently enabled. As the Digilent user manual explains, the human eye perceives the four displays as though they are continually lit, each with their own value;

5) since the Digilent board uses a single shared display bus, the others are tied off.

Here's how the game is laid out on the Digilent board.

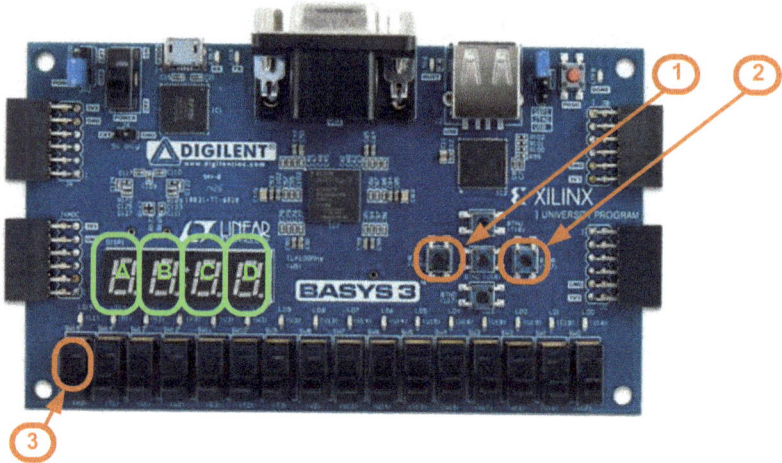

1) the "next" push-button;
2) the "affirm" push-button;
3) the game difficulty selection – off (towards the board edge) cycles each game step approximately once every two seconds; on cycles each game step approximately once a second.

A) the FIFO input segment – the segment prior to the flashing "tag" indication that must be remembered;

B) the FIFO output segment – when this matches the tagged one, pushing the "affirm" push-button scores one point. Pushing the "affirm" button otherwise, loses a point;

C) the FIFO level – never falls below four. Reaching "F" ends the game;

D) the point score.

Exercises

With the original design, as long as the user keeps pushing the "next" button, the game will continue indefinitely, regardless of missed tags. Add the ability for the game to decrement the point counter for missed tags, i.e., if the user fails to push the "affirm" button at the right time.

Increase the depth of the FIFO from 16 to 32. Since the FIFO level (how many values are currently stored) is displayed using just one 7-segment display unit, you can use the decimal period to indicate values greater than 15 (0xF).

Index

www.ingramcontent.com/pod-product-compliance
Lightning Source LLC
Chambersburg PA
CBHW071023280326
41935CB00011B/1463